CONCEPT OF MOKṢA

CONCEPT OF MOKṢA

in Advaita Vedānta with special reference to Jīvanmuktiviveka of Shri Vidyāraṇyaswāmī

By

Dr. Mrs. Vaishali Sunil Karnik

BHARATIYA KALA PRAKASHAN
DELHI (INDIA)

First Edition : **2012**

Published by **C.P. Gautam**
BHARATIYA KALA PRAKASHAN
3421-A, IInd Floor, Narang Colony, Tri Nagar,
Delhi-110035. Phone : 32967505, Fax :27393083
E-Mail : bkp3421@yahoo.co.in
Website : www.bkpbooks.com

ISBN : 978-81-8090-278-9

Printed By :
G. S. Offset Printers, Delhi.

DEDICATED

To

My Grandmother

SMT. SATYABHAMA PRADHAN

CONTENTS

Chapter		Page No.
	Foreword	*(ix)*
	Preface	*(xiii)*

PART – I

I.	Introduction	1-21
II.	Concept of Mokṣa in different schools of Indian Philosophy : (i) Cārvāka (ii) Buddhism (iii) Jainism (iv) Nyāya-Vaiśeṣika (v) Sāṁkhya-Yoga (vi) Pūrva-Mimāṁsā	22-61
III	Concept of Mokṣa in :— (i) Vaiṣṇava Vedāntins (ii) Śaivas (iii) Śāktas (iv) Prasthānatayī	62-96
IV.	Concept of Mokṣa according to Advaita-Vedānta (Uttara-Mimāṁsā) (i) Pre-Śaṅkara Advaitins (ii) Śaṅkara (iii) Post- Śaṅkara (Pre-Vidyāraṇya) Advaitins	97-127

PART–II

V. Shrī Vidyāraṇyaswāmī
–his life, date and works 128-142

VI. Traditional proofs in support of the possibility of Jīvanmukti given in 'Jīvanmuktiviveka' 143-149

VII. Nature of Jīvanmukti (Liberatīon whīle Līnīng) and Videhamukti (Liberatīon after Death) 150-158

VIII. Means to Jīvanmukti 159-181
(i) Vāsanākṣaya (Annihilatīon ofImpressions)
(ii) Manonāśa and
(iii) Knowledge

IX. Dissolution of Mind 182-194

X. The Purpose of the attainment of Jīvanmukti 195-202

XI. A Person Liberated during life (A Jīvanmukta) 203-208

Conclusion 209-215

Bibliography 216-220

Glossary 221-228

Index 229-240

FOREWORD

I feel great pleasure in writing a forword to 'The Concept of Mokṣa in Advaita Vedānta with special reference to Jīvanmuktiviveka of Sri Vidyāraṇya' by Dr. Vaishali Karnik. As a guide, I had the occasion to go through this work carefully. Certainly this is a new contribution to field of knowledge.

Shri Vidyāraṇya is one of the most important Advaita Philosophers of fourteenth century A.D. He was a great genius and versatile scholar. His contribution to Advaita Vedānta is remarkable. Though lots of research works have been done on Vidyāraṇya's other works, so far no independent and detailed study on his important work 'Jīvanmuktiviveka' is being done. Dr. Karnik has made detailed textual study on the work and presented its essence very systematically. It is certainly first of its kind in the field of Advaita philosophy.

Mokṣa or liberation according to Advaita is realization of identity between the individual self and the universal Self, the Brahman. Śaṅkara rightly points out that it is becoming one with Brahman [Brahma bhāvasca mokṣah]. The release is not something which is to be newly achieved. Release [Mokṣa] is the eternal nature of the Self. Only this Supreme fact remains unrealized because of nescience. This nescience is removable by right knowledge [Vijñānanirasya]. Religious actions only purify one's mind and it is only right knowledge leads to realization of Atman or Self. Once this ignorance is removed, one is released from bondage. This release or this state is possible in this very embodied life. This is known as state of

Jīvanmukti-living liberation or liberation while living. Final release is obtained after the death of the body [Videhamukti]. This Jīvanmukti concept which has basis in the Upaniṣads is systematically developed by the Advaita Vedāntins. Possibility of living liberation is accepted by the Buddhists, Jainas and Advaita Vedāntins in Indian philosophical field. In Buddhism the realized person in embodied state is called Arhat [in Hīnayāna] and Bodhisattva [in Mahayana], Kevali in Jainism and Jīvanmukta in Advaita Vedānta. Jīvanmuktiviveka of Vidyāraṅya is milestone in providing various proofs for the possibility of Jīvanmukti and the life of Jīvanmukta.

The author Dr. Vaishali Karnik, in this book, has done commendable job, giving brief survey of concept of Mokṣa in Indian philosophy and Pre-Śaṅkara, Śaṅkara and post-Śaṅkara Advaita Vedānta up to Vidvaranva's period [Chapters. II-III]. The fourth chapter is dedicated to describe the life and Date of Vidyāraṇya along with vast literature produced by this great Advaita stalwart. Fifth chapter is very important one in which proofs for the possibility of Jīvanmukti are discussed in detail. Sri Vidyāraṅya gives his own proofs, in addition, to support his views; he gives proofs found in Upaniṣads and writings of earlier Advaita Vedāntins. In the sixth chapter author discusses about nature of Jīvanmukti and Videhamukti given in the text. Means of Jīvanmukti is discussed in chapter seven. Vidyāraṅya mentions threefold means of liberation while living viz., Vāsanakṣaya, Manonāśa and Jñāna. This is a unique contribution made by Vidyāraṇya to obtain Jīvanmukti. This threefold means of liberation is not found in any other works of Advaita Vedānta. The learned author analysis all the three means of liberation in detail based on original text. Chapters eight and nine deal with purpose of attainment of Jīvanmukti and the state of Jīvanmukta called Paramahaṁsa-Yogi, respectively. Chapter ten describes the concept of Mokṣa in Vaiṣṇava Vedānta especially in Visiśtādvaita of Rāmānuja and Dvaita of Mādhvacarya. Vaiṣṇava Vedāntins accept only Videhamukti

[Liberation after death] and reject possiblity of Jīvanmukti [Liberation while living]. In the conclusion, the author authentically presents a comparative study of development of concept of Jīvanmukti within the Advaita Vedānta, which is a praiseworthy attempt.

This book displays the author's scholarship and insight in the subject. The subject matter is systematically presented in a lucid manner. The subject matter of the book has its own significance in the field of Indian philosophy. This is certainly scholarly work on the subject and I have no hesitation in recommending this work to the readers and students of Philosophy.

Prof. Dr. Yajneshwar S. Shastri

Director,

University School of Psychology, Education & Philosophy, Gujarat University, Ahmedabad. India.

PREFACE

Ādi Śaṅkarācārya, a great philosopher, a saint, a mystic, a religions reformer, a poet, a savant, a stiff and intrepid debater, has always fascinated me. His Advaita Vedānta is a system of great speculative daring and logical subtalty. His philosophy stands forth complete needing neither a before nor after. It has a self-justifying wholeness. He tried to bring back the age from the Puraṇas to the mystic truth of the Upaniṣad. The Advaita Vedānta alone, in the opinion of Śaṅkara, could do justice to the truth of the conflicting creeds and so he wrote all his works with the one purpose of helping the individual to the realization of the identity of his soul with Brahman, which is the means of liberation, from the world. He was in the true sense Śaṅkara i.e. bestower of happiness.

Śaṅkara's māyāwada, his concept of Brahman, illusoriness of the world, mokṣa and jīvanmukti, all are unique in the real sense. The knower of Brahman i.e. a brahmavid is a jīvanmukta. For him the world is a mirage in the desert. A jīvanmukta (a soul liberated while living) is a living God on earth, an immortal among mortals, a perfect among imperfects.

Śaṅkara's concept of jīvanmukti has been elaborately discussed in the "Jīvanmuktiviveka" by Shri Vidyāraṇyaswāmī. Vidyāraṇya also was jīvanmukta like Śaṅkara. So I was fascinated by these two concepts and chose mokṣa and Jīvanmukti for my research work. In no other religion we find these two concepts. My thesis entitled 'The concept of Mokṣa in Advaita Vedānta with special reference to 'Jīvanmuktiviveka' of Shri Vidyāraṇyaswāmī' was writen under the able guidance of Prof. Dr. Y.S. Shastri, who is honoured as one of the best scholars in the world, and is the Director of the Dept. of Philosophy of Gujarat University. The thesis was submitted to Gujarat University and was approved for the degree of Ph.D.in 1998.

This work in the book form is slightly revised version. I express my deep sense of gratitude to my guide Prof. Dr. Y.S. Shastri, who helped and encouraged me in completing this work. I owe special thanks to my friend Dr. Mrs. Sunanda Shastri, because of whom my thesis could take the book form. I also owe thanks to my dear husband Sri Sunil and sons Chi. Samir and Chi. Himanshu.

11th Sept 2010
Ganesh Chaturthi **Dr. Mrs. Vaishali Karnik**
Bhadrapad
Shake 1932

Chapter - I

INTRODUCTION

MOKṢA :

दुर्लभं त्रयमेवैतद् देवानुग्रहहेतुकम्।
मनुष्यत्वं मुमुक्षत्वं महापुरुषसंश्रयः।। (विवेक चूडामणि ३)

These three so difficult to attain, are acquired only by the kindness of Gods—humanity, desire of emancipation and the guidance of (spiritually) great men. So desire of emancipation in one's mind itself is God's gift.

Hinduism has recognised four puruṣārthas (i.e. goals) of human life, they are :—

(a) Dharma—righteousness, duty, moral order etc.

(b) Artha—Wealth and prosperity.

(c) Kāma—Desires and gratification.

(d) Mokṣa—final release or liberation. Vedānta states that Mokṣa is a state of Supreme bliss. Mokṣa is the highest spiritual value for Hindus.[1] Indian philosophy is the philosophy of values. Mokṣa is the highest value, on realising which nothing remains to be realised. It is the cessation of birth and death and all kinds of pain and at the same time it is the realization of one's own self i.e. Ātmajñāna.

The highest purpose of human life is to realize one's own nature and breath away from recurrence of births and deaths. Death is not an end of life, it is the beginning of new life. Souls cast off its worn out body and enters new one to enjoy fruits of

1. Foundation of Hinduism : Dr. Y.S. Shastri, P.115.

its own deeds, good or bad. Mokṣa is the absolute cessation of all sufferings.[2] Mokṣa means liberation from all worldly entanglements, ultimate goal of human life.

Hinduism has probed deep into the mistery of human life and come to the conclusion that life is not just period between birth and death. The Ātman is indestructible, but it goes through various cycles of birth and death on account of ignorance and its own Karma. Mokṣa is to be free from the cycles of birth and death, to be free from all the karmas. Ignorance is bondage and knowledge is Mokṣa.

The aim of philosophical inquiry is mainly the attainment of liberation, which is also called as mukti, mokṣa, amṛtatva, niḥśreyas, self-realization. Knowledge is the means to the end of self-realization. It is Brahmānubhava, being one with Brahman, experiencing Brahman.

The concept of 'Mokṣa' is an unique concept in Indian Philosophy. 'Mokṣa' is usually translated into English as liberation, salvation or release but in fact none of these words reveal the perfect meaning of it. The conception of salvation or release in Western Theology implies that the soul has fallen in the death of the world at some particular time due to certain sin, and we have to release it from the world, Indian conception of Mokṣa implies that the soul is eternally free, but it has wrongly identified itself with certain extraneous elements and we have to release its true nature. In this lies Mokṣa.[3]

A Philosophical system in India is called a Mokṣa-śāstra, the science of liberation. The aim of all the systems in Indian Philosophy is to show the mean for the attainment of liberation. The Indian Philosophy is an art of living a well as science of thinking.[4]

Mokṣa has been rightly called by Dr. S.N. Dasgupta as 'the pivot on which all the systems of Indian Philosophy revolve'.

2. Ibid.—P-117.
3. Comp. study of the concept of mokṣa. Preface A.K. Lad.
4. Ibid. : Introduction.

He writes——'The doctrine of mukti and karma are the two fixed postulates which Hindu Philosophy could not disvow even in its highest soarings. These two doctrines taken alongwith their necessary corollaries sum up all the important peculiarities of the Hindu mind as it manifests itself in philosophy.[5]

The 'Mokṣa' (sanskrit. Mukti, pali vimutti, or vimokha) comes out of the Sanskṛt root 'muc (+ ktin) meaning 'to loose set free', 'to let go', 'discharge', 'release', 'liberate', 'deliver', etc. These are all identical in primary meaning with our 'freedom', 'deliverance', 'emancipation', 'release', 'final deliverance' or absolution of the soul from metempsychosis.[6]

The concept of Mokṣa has two aspects, viz. a negative and a positive one. Negatively Mokṣa means 'freedom from sin', 'freedom from fear', 'freedom from grief', 'freedom from desire', 'freedom from body, decay, birth, death, attachment etc.' Overall, it means the freedom from all the effects of man's actions or karmas that bind him to this lower plane of existence.

The positive aspect of Mokṣa is that it is a state of Supreme Bliss. A realīzsed soul attains the culminatin of happiness. The satisfaction that results from sense object is dependent, relative and limited, whereas the satisfaction concequent upon Brahman knowledge is without any limitation. Being in the nector of knowledge he is fully, contended. Knowledge itself is Mokṣa.[7]

Dr. Rādhākriṣṇan holds that Mokṣa is a matter of direct realization of something which is existent from eternity, though it is hidden from our view. When the limitations are removed, the soul is liberated. It remains where it is what it is and eternally was, the first principle of all things. It is the peace that the world can never give, nor take away, the supreme and the only blessedness.[8]

Acquiring Liberation (Mokṣāvastā) is becoming one with

5. Yoga Philosophy : S.N. Dasgupta., p. 316, 10.
6. Ad. concept of J.M. : L.K.L. Shrivāstava, p. 9.
7. Ibid., p. 10.
8. Indian Philosophy : Dr. S. Rādhākriṣṇan., p. 636-637 vol. II.

Brahman. Brahman which is real in absolute sense. It is immutable, eternal, all penetrating like ākāśa, without any change, all satisfying, undivided. It is self-lumineous beyond time and space. This incorporeal is called Mokṣa.

Mokṣa is the realization of supreme bliss, which is the essential nature of self, Gold shines only when it is freed from impurities, which affect it. The stars shine only when the clouds are removed. In the same way the true naure of the soul stands self-revealed only when (Avidya) nesience vanishes. When ignorance is removed, the nature of the self as unmixed bliss manifests and shines itself.

The realization of the truth does not mean the abolition of plurality but only the removal of the sense of plurality. It is an insight, which changes the face of the world, and 'makes all things new.' This insight, this changed attitude to life and its happenings is not so much a condition of Mokṣa itself. The unending procession of the world will go on through its ups and downs, but the liberted man's attachment to it is over. On the attainment of Mokṣa nothing happens to the world, but only our views of it alter. Mokṣa is thus not the dissolution of the world but only the disappearance of the false outlook.[9]

Vedas Believe 'Heaven' as the Highest Goal of Life

The history of philosophy begins with the Vedic Hymns. The Vedas are supposed to have divine origin. They are a apauruṣeya or non-human. They are not intentionally composed by the ancient sages but they were revealed to them in their intuitive experiences.

Out of four Vedas'—the Ṛgveda and the Sāmaveda, the Yajurveda and Atharvaveda—The Ṛgveda is supposed to be the basis of philosophical thought. So Ṛgveda is the chief Veda. The inspired songs which the Āryans brought with them from their earlier home into India as their most precious possession were collected, it is generally held, in response to a prompting

9. Ibid. : p. 637.

to treasure them up which arose when the Āryans met with large numbers of worshippers of other Gods in their new country. The Ṛgveda is that collection.[10]

The thinkers of the Vedic period personified the natural agencies and things are ascribed to them desires and will, good and bad with which they could affect human life. Naturally out of the awe and a desire for security, the powers came to be looked up as deties. They attributed power to everything in the world that affected them or impressed them with certain powers in someway. They vaguely suspected some 'energising principle' actually investing all the things of the world with energy, and itself remaining imperceptible. It is found after a careful scrutiny of the hymns of Ṛgveda that 'manas' stand on equal par and denote a factor in man which is subtler and is different from his gross body.[11] The word 'manas' has been used with many shades of meaning. It is generally supposed to be something which is intelligent, active source of vitality, the doer of action, abiding and the owning prinicple.

Vedic hymns exibit a kind of primitive faith in polytheism as the various objects of the world are thought of separately and each of them is worshipped as a separate deity. In the beginning, the Vedic thinkers were not aware of a unitary principle working behind the manifold, but gradually the philosophic thought progressed and the new idea of the Ṛta begins to appear. Ṛta is a single power that maintains the regularity and order of the processes of nature. It also maintains the moral order of the universe and remains at the heart of the reality.[12]

The Śatapatha Brāmana holds that every man has three births. He receives first birth[13] from his parents; the second one at the time of wearing of the sacred thread; and the third

---

10. Ibid., p. 64.
11. Atman and Mokṣa : Dr. G.N. Joshi, p. 5-6.
12. Ibid., p. 11.
13. Ibid., p. 14.

after death. The mention of the third birth of the individual after death clearly shows that the Vedic thinkers believed in the survival of some non-physical or immaterial element after the destruction of physical body i.e. soul.[14] They also believed that the soul after abandoning the previous body assumes another body.

The dead persons rise to the land of Yama after death. Yama, the first royal man had gone ahead and had secured for himself and his successors, a territory in heaven. Ṛgvedic seers hold that all dead persons go to the land of Yama i.e. heaven. All kinds of imperfections are absent in it and it is full of happiness. The soul is not destroyed along with the body but remain even after dath. The concept of heaven is similar to a happy and comfortable life on earth. There is eternal light and there all persons are equal and weak are no longer subject to pay tribute to the strong. In this heaven, the sound of the flute and song is heard.

Thus the Vedic thinkers positively deemed at reaching the heaven to live a fuller and more perfect life. In this lies their Mokṣa. Their Mokṣa has a positive connotation, The ideal Mokṣa is an elaboration of a birth and happy earthly life. The end and aim of human life is the attainment of heaven, where perfect life can be employed. Immortality of the soul is combined with positive enjoyment.

Upaniṣadic Concept of Mokṣa

The Upaniṣads form the concluding portions of the Veda, and are called, therefore, the Veda-anta, or the end of the Veda, which suggest that they contain the essence of the Vedic teachings.

'The word Upaniṣad comes from 'Upa-ni-sad' which means 'sitting down near'. It means sitting down near the teacher to receive the instructions. It gradually came to mean what we received from the teacher, a sort of secret docrine, sometimes

---

14- Ibid., p. 16

it is made to mean what enables us to destroy error, and approach truth. Śaṅkara in his introduction to the Taittirīya Upaniṣad says, 'Knowledge of Brahman is called Upaniṣad because in the case of those who devote themselves to it, the bonds of conception, birth, decay, etc., become unloosed, or because it destroys them altogether, or because it leads the pupil very near to Brahman, or because therein the highest God is seated."[15]

Upaniṣads hold that mokṣa is highest condition of disintegration of individuality, a giving up of selfish isolation and not mere death.

"Mokṣa is the highest state of rapture and ecstasy, a condition of (ānand), where the creature as creature is abolished, but becomes one with the creator, or more accurately realises his oneness with Him. We can not describe this perfection adequately. We use symbols. The nature of eternal life is a condition of bliss (ānanda) or freedom, a state of joyous expansion of the soul, where heaven and earth are felt to flow together."[16]

Muṇḍaka Upaniṣad says that the flowing rivers loose their name and form as soon as they disappear in the sea, even so a wise man, freed from name and form, goes to the divine person who is beyond all" (iii.2.8)

'Praṇava is the bow, the Ātman is the arrow, and the Brahman is said to be its mark. It should be hit by one who is self-collected, and that which hits becomes like the arrow, one with the mark, that is Brahman.' (ii.2.2)

'All these become one with the highest imperishable Brahman' (iii.2.7)

'He becomes omniscient and becomes all.' (iv.10)

'Having attained Him, the seers content with their knowledge, their purpose accomplished, free from all desires and with full composure, having attained the all pervanding

15. Indian Phi. : Dr. S. Rādhākṛiṣṇan, p. 137/138.
16. Ibid., p. 237.

Ātman on all sides, ever concentrating their minds, enter into everything. (iii.2.5)

Individual soul is the real doer of actions. It supplies energy to the mind, buddhi etc. It is the initiator and sustainer. Though a seeker of mokṣa does the actions, its soul is not bound by them. It perceives and acts through sense organs, but itself remains aloof from all of them. So the results of actions do not stick to it as a drop of water does not to the lotus leaf.

Yājñavalkya says in Muṇḍaka Upaniṣads, 'He is the unseen seer, the unheard hearer, the unthought thinker, the ununderstood understander. Other than he there is no understander. He is your soul the inner, controller, the immortal.' (iii.7.23)

'One who knows this Brahman, becomes Brahman only, he crosses griefs and all sins, becomes free from the knots of all cells, attains immortality.' (iii. 29)

Thus the liberated soul eperiences oneness with Brahman. It is the experience of non-duality. It is only the discovery of one's innermost self that was lost due to ignorance. It is experiencing Brahman, becoming one with Brahman which is called as 'Brahmānubhava'. The self which was falsely understood under various kinds of adjunct (Upādhis) is rightly understood in Mokṣāvasthā. (State of liberation).

In Bṛdāraṇyaka Upanisad Sage Yajñāvalkya says—'The self is dearer than a son, is dearer than wealth, is dearer than all else, since the self-alone is nearest...One should reverence the self alone as dear.' (i.4.8)

Thus the nearest thing that an individual ever has, is his soul. All our activities pertain to the soul-whatever we feel dear to us, is not dear for the sake of the thing, but for the sake of self. Again moral behaviour of an individual self plays an important role in attaining Mokṣa. Sage Yajñāvalkya says, 'This self, then as his conduct and behaviour has been, so does he become. He whose works have been good become good, he whose works have been evil, become evil. By holy works, he becomes holy. By sinful works, sinful. It is for all this reason, that they say that a person consists merely of desires, as his desire is so is his will, as his will, so his work, as his work, so his evolution (4.4.5) So by his pure deeds a self-attains his liberation. Mokṣāvasthā is perceived and experienced just by mind.

Katha states that—'The self can not be gained by understanding nor by much learning nor by much hearing and preaching. He whom the self-chooses, by him, the self can be gained.' (i.2.23)

'He who has perceived which is without sound, without touch, without form, without decay, without taste, eternal, without smell, without beginning, without end, the great and unchangable, is freed from the jaws of death.' (i.3.15)

So the liberated soul is looked upon as a perfected individual with a status of absolute equality with the Supreme Soul. (Muṇḍaka iii.1.3)

According to the Chāndogya immortality is lifting oneself up to the region of the deity (ii.22). Muṇḍaka holds it to be the companionship with God (iii.2.6) Thus Mokṣāvasthā is to become God. It is a state of activity, full of freedom and perfection. It is a state of pure divine life. As the ray of the Sun is lost in the Sun, as the waves of the ocean are lost in the ocean, as the notes of music are lost in one harmony, so the individual self is lost in Supreme Self. The individual self becomes one with all and lives a life in unity with God. Individual becomes one with the highest. The individual self merges into Supreme Self and itself becomes highest imperishable Brahman. He becomes omniscient and enters into all and becomes all. 'The liberated soul feels his oneness with God so intensely that he calls himself the creator of the world. 'I am food', I am the food-eater I am the subject, I am the object I am the two togther, I am the first born, the destroyer of the world also, I am the sun like light, I am the centre of the world, of immortal God. (Taitt. iii)

Yajñāvalkya explains Mokṣāvasthā in the words—'As a lump of salt which is thrown into the water dissolves and cannot be gathered up again, but wherever water is drawn, it is salty, so truely is it with this great being, the endless, the unilimted, the fullness of knowledge, from these beings it came into view and with them it vanishes, there is no consciousness after death. (Muṇḍaka iii.1.3)

Thus the Upaniṣadic thinkers held that the aim of philosophical inquiry was the attainment of Mokṣa. Mokṣa is mukti or liberation. It is amṛtatva, self-realization, liberation, niḥśreyas. It is nothing but becoming one with the whole, merging of an individual self into the Supreme Self.

Types of Mokṣa or Mukti

All the Indian philosophical systems lay much stress on mukti as the end of life because unlike western philosophy (which arose from the sense of wonder about the whence and wither of life), Indian philosophy begins with a reflective examination of the state of human life as suffering and the means to get freedom from this. Many, particularly ascetic authors, try to show that the wordly existence of man, is full of suffering; that much of it is a fight against and fight from this suffering. For this purpose, they mention the physical, psychical and supernatural sufferings that a being must endure from birth to death, and they endeavour to show that even that which gives pleausre is in reality only pain. All the religious thinkers of India are of one accord that there is certainly a way to be free from the sufferings, pain and evils of life, there is a way to break the chain of births and deaths, repeated existences and attain the true happiness forever. Thus they suggest the goal as the attainment of Mokṣa or mukti. Mokṣa is nothing but ātmasākṣātkāra (Self-Realization),[17] apavarga of Nyāya—Vaiśeṣ ika, Kaivalya (jainism and Sānkhya) and Nirvāṇa or bodhi (buddhism).

Indian Philosophy Recognises various Type of Muktis :

'Śabdakalpadruma enumerates its five types viz. Sārṣti, Sālokya, Sārūpya Sāyujya, and Nirvāṇa. It further clarifies that the first four types of mukti are desired by Vaiṣṇavas and the

17. Advaitic concept of Jīvanmukti: L.K.L. Shrivāstava, p. 9.

fifth one is aimed at by other Sādhakas. The Sūtta-samhitā, in its section on deliverance, makes a broad distinction between delivergence achieved through knowledge e.g. Jñānaphalabhuta and that produced by action e.g. Karmaphalabhūta. Again it divides the latter into four types viz. Sālokya, Sāmīpya, Sārupya and Sāyujya what Saṁhitā calls jñānaphalaphūta makes the cardinal point of Advāita Vedānta. By jñānapalabhūta it may mean Advaitic types of mukti viz. Jīvanmukti, Videhamukti, Kramamukti and Sadyomukti, because all of them result from the attainment of knowledge.[18]

Another basis of classifying mukti into various types is the problem of man's association or dissociation with the body at the time of attaining liberation. The former refers to Jīvanmukti and the later to Videhamukti.

A third distinction can be made on the basis of process involved in release and the efficacy of different disciplines. Under this category we find, generally Sadyomukti and Kramamukti.

Further mukti may be classified on the basis of the nature of perfection attained and here we may enumerate all the four types of Vaiṣṇavic muktis as stated above.

A broad distinction of mukti may be made on the basis of self-realization vs. God realization. The matter in question, here, is whether mukti consists of self-realization.Advaita Vedānta Sāṁkhya-yoga, Jainism etc. accept the first alternative, where as systems like Viśiṣṭhādvaita, Śaiva-siddhānta, Kaśmir Śaivism etc. bring another entity called God whose association contributes mokṣa.

Still another classification is related to the attainment of liberation by individual and liberation of mankind as a whole—Shri Aurobindo conceives two types of mukti viz. individual; and cosmic (sarva) mukti. The latter one is attributed greater importance.

---

18. Ibid., p. 11.

Dr. S. Rādhākriṣṇan is also an advocate of universal salvation. Appayya Dikshit, in his Siddhānta-leśa-sangraha argues in favour of Sarvamukti and basing his view on Pratibimbavāda, concludes that a liberated Jīva attains a state of Iśvara and waits for the final absorption into the Absolute till the moment of Sarvamukti i.e. universal emancipation.[19]

Vaisṇavism recognises four types of muktis such as Sālokya, Sāmipya, Sārūpya and Sāyujya. Sālokya mukti refers to the sharing by the devotee of the same sphere as is occupied by the deity.

Sāmipya is the proximity to the goal of liberation won by those who perform the duties of their respective stations in life.

Sārūpya is the sameness of four with the deities won by the devotees and the last Sāyujya aims at the unity of devotee with the deity worshipped.

Advaita Vedānta recognises four types of mukti, viz. Sadyomukti i.e. immediate deliverance, Kramamukti i.e. gradual deliverance, Jīvanmukti i.e. liberation while living, and Videhamukti i.e. deliverance after death. Let us see these muktis in detail.

Sadyomukti

Sadyo means immediate and mukti means deliverance. It denotes that the aspirant passes to final liberation at once with the down of the self-knowledge. Śaṅkara says that immediate deliverance is attained by those renunciates who have established themselves in Saṁyagadarsana,[20] i.e. right vision. Śaṅkara refers here those whose happiness is within themselves, who delight within themselves, whose light is within, those who are yogins and have become one with Brahman, their sins have all dwindled away, their doubts have all been solved, they have perfect sway over themselves and are intent on promoting the welfare of all living beings. They are free from lust and anger

19. Ibid., p. 12.
20. The cocept of mukti in Ad. Vedānta : A.G.K. Warrier, p. 476.

they have known the self and are, therefore, delivered both this life and hereafter. Logically Sadyomukti must be identical with videhamukti. Both may be alike defined as Jñānasamakālamukti-deliverance, simultaneous with the knowledge that saves.

Kramamukti

Kramamukti means gradual deliverance. This type of mukti falls to lot of those who worship Saguṇa Brahman and are possessers of lower knowledge. The worshippers of Saguṇa Brahman pass initially to the world of Hiraṇyagarbha by the way of Gods or Devayāna. It is the subtle body and not the 'Ātman' of the sādhaka moves to Devayāna. During this journey, along the path of Gods, the sādhakas are helped and directed by the deities assigned to different stations. After that a non-human guide conducts them to Brahman i.e. Saguṇa Brahman. The image worshppers, in the realm of Kārya Brahma, 'enjoy a state of quasi-deliverance consisting in hightened power of enjoyment.' They do not return to the lower region of Saṁsāra (world) any more. In due course of time they attain immortality.

Śaṇkara accepts this type of mukti because he has to rely upon Śruti. Śruti recognises such type of mukti. Śaṅkarā clearly says that—"we have to recognise deliverance by stages because, the śruti teaches movement and non-return. Moreover he seems to stick to the ancient faith which says that there is a heaven beyond this earth, the abode of virtuous and wise souls. Advaita Vedānta subscribes faith to kramamukti, yet even here unqualified deliverance can only be attained by the attainment of the right knowledge of the absolute. So Advaita Vedānta does not deviate from its firm hold that mukti is attained by knowledge.[21]

Jīvanmukti :

Jīvanmukti means liberation while living, liberation before

21. Ad. concept of J.M. : L.K.L. Shrivāstava, p. 14.

death, liberation while you have bodily existence. Even if you have bodily existence you get liberated from all obligations to future births. So Jīvanmukti means the final liberation in the present state of life.[22] When the bondage of the living being which consists in those functions of mind which are characterised by feeling of pleasure and pain concomittant with action and enjoyment, is removed within this life, the state is called Jīvanmukti. So term 'Jīvanmukti' is used to discribe the state of a continued life in the body on the part of a knower of Brahman. He becomes the Brahman after knowing it. A Jīvanmukta is a spiritual leader of the society. His action tend towards the spiritual uplift of the whole human race. He is a living God on earth. He is a Yogi of highest cadre (Paramahaṁsa-Yogin). He is entitled as mahātman, a mahāpuruṣa. He is sthitaprajñā in Bhagavadgītā. He is also called as ativarṇāśrami. An ativarṇāśrami is beyond varṇa and āśrama who knows his self as God, from hearing the great texts of the Vedānta. He is also called as bhakta, a Brāhmaṇa, a guṇātīta, such a enlightened yogin is liberated while living (jīvanmukta). Thus—

> जीवन्मुक्तस्तत्त्वविद्यः शास्त्रयुक्त्यनुभूतिभिः।
> तस्य प्राणादिनामान्तं जगत्स्त्वात्मन उद्गतम्।। अनुभूतिप्रकाश iv.84

The liberated and yet alive is the one who has realised the self on the basis of scripture, reasoning, and experience. For him, the world beginning with vital air and ending with name springs from the self.

Videhamukti

Jīvanmukti is the penultimate stage and videhamukti is the last stage of a liberated soul. Finally liberated sould is called videhamukta; i.e. the bodiless liberated soul. He is dissociated from psycho-physical organism. The word 'videha' itself means one who has no relation with the body or who is detached from the body. He is detached from the gross body as well as subtle body.

Of the various types of liberations (mukti) discussion

22. Ibid, p. 15.

above, Jīvanmukti holds a supreme position. If knowledge is the sole means to attain liberation, it can be attained here and now, and the mukti, thus attained in one's life time, is positive and real. The concept of Jīvanmukti is logical and variable in one's experience. None can deny the records of mystic experience of the sages and saints who have attained self-realization. Their experience is characterised by a certain 'givenness' which makes it authentic and genuine. It is not a matter of argument.[23]

Path of Liberation : *(means of Mokṣa)*

Advaita Vedānta believes that it is knowledge of the self-alone that leads to Mokṣa. Śaṇkara gives supreme importance to the higher knowledge of the self or Brahman as the only means of the attainment of Mokṣa. He was essentially a jñānavadin. He believed that avidyā is the real cause of bondage. As soon as the real knowledge of Brahman dawns false knowledge disappears. Knowledge does not mean mere study of the Vedas, it is in a true sense experiencing 'अहं ब्रह्मास्मि' 'सर्वं खलु इदं ब्रह्म' 'तत्त्वमसि।' etc.

न वेदपाठबाहुल्यान्न बहुश्रुतितस्तया।
आत्मा लभ्योऽथ देवो यं वृणुते तेन लभ्यते।।[२४] vi.84

The self is not attained by an abundant study of the Veda,. nor by much hearing. It is attained by one whom the God blesses.

So it is a grace of God resulting from bhakti or loving devotion towards Him that you wish to attain Mokṣa. Thus to earn the grace of God devotion (bhakti) is the means.

सत्यं तपो ब्रह्मचर्यं विद्याधिकृतिकारणम।
मुख्याधिकारी विद्यायांवरिष्ठो ब्रह्माविद्भवेत्।।[२५] vi.79

Truth speaking, control of sense and mind, continence make one competent to attain the knowledge of Brahman. He who is the most competent one to attain the knowledge (by

23. Ad. concept of J.M. : L.K.L. Shrivastava, p. 19
24. Anubhūtiprakāśa of Vidyāraṇya, p. 207 : Godabarisha Mishra.
25. Ibid., p. 206

possessing the above prerequisites) becomes the best among the knowers of the truth[25]

श्रवणं मननं ध्यानं यस्य न प्रबलं भवेत्।
प्रमादो वास्ति विषयासक्त्यात्मा न स बुध्यते।।[२६] vi.87

- The self is not realised by one in whom vedāntic study, reflection and meditation are not intense, or there is inattentiveness owing to attachment towards worldly objects.[26]

Śravaṇa, manan and nididhyāsana are also necessary to attain Mokṣa. The identity of jīva and Brahman is revealed through the mahāvākyas. By the means of mahavākyas the discovery of their meaning (viz. the identity of jīva and Brahman) is called inquiry i.e. Śravaṇa. On the other hand, the understanding of the possibility of such identity by reasoning, is reflection.

Śravaṇa is an inquiry to ascertain and discover the exact import of statements in the Śruti. It consists of exposure to the teaching by a competent master, wherein the real nature of oneself is unfolded as per the Śruti. It is not just listening, but doing so with a spirit of investigation and ascertainment leading to the discovery of oneself. For Śravaṇa to be effective, the person should first acquire (Sādhanacatuṣtyasaṃpatti) and the purity of mind accompanied by withdrawal from its hectic preoccupations. The subject matter is not objective and so it is difficult, making repeated listening (Śravaṇam) indispensible until the clarity of knowledge is gained.[27]

Proper undertsanding of the rationality of such discovery is the next stage and it is contemplation (manan). There are enough reasons to entertain doubts about the teachings from Śruti. Our perceptions convinence us about the reality of the world, whereas Vedānta emphasises without any compromise, that nirvikalpa Brahman alone exists and that there is no room for the existence of any real thing other than Brahman. Jīva and Īśvara also appear to be totally different. How do we

26. Ibid., p. 208
27. Vedānta Pañcadaśi : Swami Śuddhabodhananda, p. 132.

reconcile these contraditions between our immediate experiences and the declarations in the Śruti? Therefore it is imperative that these teachings are tested on the touchstones of reasoning to clear the doubts This reasoning process is called manana. With the help of reasoning, the clear understanding that the ascertained meaning of Śruti is possible. This is manana or reflection. Thus manan is meant to clear the doubts regarding the knowledge gained from Śruti.[28]

Nididhyāsana is mental operation helpful to turn away the mind from external objects in order to maintain the continuity of the knowledge in the form of 'I am Brahman,' that arises from śravaṇa and manana. 'The steady abidance of the mind that is fixed in the thing (i.e. Brahman) about which the doubts are dispelled by śravaṇa and manana is called nididhyasana.'[29]

Mind is fixed in the appreciation of Brahman to the exclusion of all other dissimilar thoughts, to enable it to get absorbed in it. This is to be accomplished by training the mind to take to the one and the only one preoccupatin of entertaining a vritti that appreciates Brahman.

The mind when left on its own, occupies itself with hundreds of different preoccupations by force of habit, if not any specific purpose. It does not find any leisure to appreciate one's real nature. Volitionally or habitually we persue various things whereby the mind remains constantly proccupied with everything else but nididhyāsana. Basically nature of mind is unruly.

Nididyāsana is a continuous and repetitive persuit. The mind fixed in the knowledge of Brahman steadily abides in it to the exclusion of the other dissimilar thoughts, after sufficient practice. This is called 'ekatānatvam' i.e. a continuous flow of the same type of state (vrtti). Such a steady abidance of the mind in the appreciation of Brahman which is really oneself is called nididhyāsana.[30]

---

28. Ibid., p. 133.
29. Ibid., p. 136.
30. Ibid., p. 137.

Nididhyāsana is ncessary until the erroneous identification with body is totally destroyed. All continue to have identification with the body habitually even after gaining the sef-knowledge, exception being very rare in the entire creation. Let us not imagine ourselves to be one of them. The knowledge that—'I am Brahman' must become as natural for ourselves as the gross body is 'I' for an ignored person.Contemplation being the appreciation of my true nature is limitless Ānanāda.[31] A seeker of Mokṣa must have pure mind and body. No doubt self is self-illumination, but if it is enveloped by a dark medium, it will not be visible. A mumūkṣu must have clean and transperant physical structure, must remove all the impurities and make the body and mind clean so that the soul's brilliance become visible.

Importance of ethical life also cannot be neglected. A mumukṣu must be ethically good person; Avaita Vedānta recognises need for and importance of ethical life in so far as it works on the phenomenal plane, it emphasizes the good and the right as against the evil and wrong and the seeker of Mokṣa to rise beyond the gross adjuncts which make the knot of bondage more and more tight. Ethically good and right is not the ultimate and to be attained. Śaṅkara emphasizes that the end is the attainment of Brahman itself by becoming one with it, a state of absolute perfection which is beyond all dualities and relativity. All our distinctions of good and evil, right and wrong, mine and thine are dissolved in the final state of mokṣa. It is not a state of morality. Morality helps the individual to become progressively free from the vanity, pride and egoism, which are falsely created by nescience. Practice of ethical virtues liberates an individual from his limiting adjuncts and helps him to widen his consciousness of the Brahman.

Śaṅkara gives importance to all these various kinds of ethically important actions and mental attitudes, but he does not thereby lose the sight of knowledge. He ultimately maintains that all these ethical activities may strengthen to accelerate the attainment of Mokṣa, but ultimately Mokṣa can be attained

31. Ibid., p. 138.

only by knowledge and knowledge alone, for Mokṣa is nothing but knowledge of the Brahman, and the knowledge of Brahman is nothing else, but being Brahman itself. It consists in the infinitisation of the finite.[32]

The study of the Vedas and scriptures is necessary for properly grasping the nature of Brahman. He rejects the sacrifices prescribed by the Śṛti. He does not admit that the mechanical performance of the Vedic rituals lead to Mokṣa, however he admit them as necessary means in so far as they help the purification of mind and buddhi.

A seeker of Mokṣa must renunciate all his desires and passions He must develope the attitude of detachment to things to make himself free from the influences of the adjuncts that create nescience. The development of 'vairāgya' or detachment and disinterestedness helps to a very great extent to make soul free from the various adjuncts which cause afflictions to it.

Śaṅkara recognises all the eightfold yogic practices like yama, niyama āsana, prāṇāyāma, pratyāhāra, dhyāna, dharaṇā and samādhi, as the means of mental and physical discipline. Self control, renunciation and resignation help the individual soul to detach itself from the worldly objects and their influences on it which cause attachment and which make the knot of bondage stronger.

Concentration and meditation also are necessary to realise one's real nature and to uproot all other false attachments to various worldly objects of enjoyment.

Śaṅkara believes that the Vedic rituals may lead to enrichment of worldly life but not to salvation. Knowledge alone can effect mokṣa. So mokṣa can be attained by knowledge and by knowledge alone.

Śaṅkara maintains that people of all casts can attain mokṣa irrespective of the sex differences. However, he recognises the traditionally sanctioned Varṅāśrama dharma and holds that the high cast people can attain it from the study of scriptues and the others by bhakti. He nevertheless allows

32. Ātman and mokṣa : Dr. G.N. Joshi, p. 506, 507.

mokṣa to any person of any caste, any sex and any race, if the knowledge of Brahman dawns upon him even without any external means.[33]

Vidyāraṇya's Jīvanmuktiviveka

Vidyāraṇyaswāmī was said to have flourished towards the fourteenth century A.D. and (A.D. 1297-1386) lived for ninety years. The name of his Pūrvāśrama was mādhavācārya i.e. his pre-renunciation name was Mādhavācārya.. He was a great teacher, a grammarian, a commentator, a logician, a versatile scholar, a foremost exponent as hereditary and chief minister of the Vijayanagar and the founder of Vijayanagar Empire. He was given a right appellation Vidyāraṇya', i.e. Forest of Learning,' was associated with the Advaitic school of thought.

The doctrine of Jīvanmukti has been very well expounded by shri Vidyāraṇya in his precious work 'Jīvanmuktiviveka'. Śaṅkara's Advaitic concept of Jīvanmukti is elaborately discussed in it in all its subtlty and depth. The 'Jīvanmukti viveka' or 'the path to liberation in life' is unique in its own way.

'Jīvanmuktiviveka' is divided into five chapters. The first one is Pramāṇ-prakaraṇa, i.e. authority bearing testimony to Jīvanmukti. The conception of Jīvanmukti is based on self-certifying experience of the saints and mystics. Vidyāraṇya himself being a saint, a sanyāsin and a jīvanmukta, advocates the ideal of Jīvanmukti., proves its possibility by quoting evidences from Śrutis, Smritis especially from Laghu-Yoga-Vāsiṣṭha. 'Jīvanmukta' has been denoted as guṇātita, a sthitaprajñā, a brāhmaṇa, a bhakta, an ativarṇāśrami etc. in different works.

The second and the third chapters deal with the vāsanākṣ aya, manonāśa and knowledge as the means to attain Jīvanmukti. The largest portion of 'Jīvanmuktiviveka' is devoted in describing and analysing the threefold means for obtaining liberation while living. The three means are destroying mental impressions, extinguishing mind and knowing the truth. These three means should be practised repeatedly and not at intervals

33. Ātman and mokṣa : Dr. G.N. Joshi, p. 505, 506, 507.

and again these three factors are internlinked and interdependent on each other and cannot be practised in isolation or one after another. Though these are the three means for attaining living liberation, Vāsanākṣaya and Manonāśa are the principle means and knowledge of the self is subordinate or the cause of the rest two means, holds Vidyāraṇya. This is the key to Jīvanmukti.

The fourth chapter discusses the purpose of attainment of Jīvanmukti. Vidyāraṇya gives a logical and consistent justification as to why one strives for the attainment of Jīvanmukti. In this chapter he enumerates five purposes to attain Jīvanmukti. They are—ज्ञानरक्षा तपो विसंवादाभाव दु:खनाश सुखानिर्भावा: सन्ति पन्च प्रयोजनानि। *i.e.*

(1) Preservation of knowledge.
(2) Practice of Penance.
(3) Absence of discard.
(4) Cessation of pain and misery.
(5) Attainment of Supreme bliss.

The last and the fifth chapter discusses renunciation of the knower or Vidvat-Saṅyāsa-prakaraṇa. It deals with the renunciation of Paramahaṁsa-Yogin i.e. Vidvat-Saṅyāsa. Author recounts Jīvanmukti way of life most clearly here in this chapter. As it forms a commentry on the Paramahaṁsa Upaniṣad, he describes how the supremly ascetic yogin or realized renuncient, interweaves yogic practice and renunciation. The discussion illustrates both the path to and life styles in Jīvanmukti.

The path of Paramahaṁsa is the most uncommon path and is very rarely to be met with in this world. Though the path is difficult to follow, its not impossible. One should not remain without striving towards the end. Hardly anyone can reach the end.

In Bhagwad Gīta Lord Kriṣṇa says :

मनुष्याणां सहस्त्रेषु कश्चिद्यतति सिध्दये।
यततामपि सिद्धानां कश्चिन्मां वेत्ति तत्त्वत:।।७.३।।

Among thousands of men, scarce one strives for perfection. Of the successful strivers, scarce one knows Me in truth.

Chapter - II

CONCEPT OF MOKṢA IN DIFFERENT SYSTEMS OF INDIAN PHILOSOPHY

Cārvāka

Bṛhaspati is said to be the founder of Cārvāka philosophy. Accounts of the Cārvāka systems are found in 'Sarvadarśana Sangraha' of Sāyaṇa Mādhava and "Ṣaddarśanasamuccaya of Haribhadra.

The Cārvāka school came into existence in the post-vedic period. The four schools namely Cārvāka system, Buddhism, Jainism and Pūrva-Mimāmsā are unanimous in denying the existence of God, simply because they can explain the world and life without the idea of God. Out of these four schools, the Cārvāka philosophy is not only atheistic, but it is also anti-theistic. The Cārvāka philosophy, Budhism and Jainism were powerful reactions offered against the Vedic and Brāhamaṇic ritualism and sacrifices. They refused to believe in the Vedas and in their divine origin. They did not believe in animal and human sacrifices and their efficacy. The Purva-Mimāmsā system, of course, believed in the validity, authority and divine origin of the Vedas but the Cārvāka philosophy offered a deadly resistance to the Vedic ritualism and sacrificalism.[1]

The Cārvāka school is known also as the Lokāyata school of philosophy. The Cārvāka are sometimes known as the Āsuras or dehātmavādins. They denied the existence of the soul as something different of the living body and superior to it. They refused to admit that the soul is an eternal and transmigratng

1. Ātman and Mokṣa : Dr. G.N. Joshi, p. 100.

entity. As a consequcne of it, they denied the existence of the life after death, heaven and rebirth.

There were two sub-schools of Cārvāka system. They were the Dhūrta-Cārvākas and the Suśikṣita-Cārvākas. Both of them were materialisti. Both of them denied the existence of the soul as a spiritual entity. They believed that the end of life is enjoyment pure and simple as long as one existed. The aim of life consisted acording to them is the highest bodily enjoyment of every sort.[2]

The Cārvāka regarded perception as the only means of valid knowledge and reject the validity of inference. They are free thinkers. They reject the autority of the Vedas and the supremacy of brāhmanas. they are naturalists and accidentalists, they reject final causes and universality of causation. They are materialists, positivists, atheists and hedonists. They identify the soul with body and deny pre-existence, future life, law of karma, haven and hell, bondage and release and the existence of God. They admit the reality of gross matter and advocate egoistic hedonism in ethics.

The Cārvākas did not recognise any supramundane existence. Death was the end of life according to them. They never believed in the self nor they believed in the continuation of life after death. 'For them death was the extinction of all life and Mokṣa was nothing but the destruction of body, which one meets inevitably.'[3]

They remained unscrupulous about the moral and immoral aspects of life because there was no question of reaping fruits of evil deeds. The goodness and badness of actions was determined according to their own thoughts. They thought that an individual must achieve maximum physical happiness. It is the aim of life. They were unrefined hedonists.

Dr. Rādhākriṣṇan describes his view as follows : Virtue is

2. Ibid., p. 101.
3. Sarvadarśanasangraha : Sāyaṇa Mādhava, p. 6.

the delusion and enjoyment is the only reality. Death is the end of life. There was distrust in everything good, high, pure, compassionate. The theory stands for sensualism, selfishness and the gross affirmation of the loud will. There is no need to control passion and instinct, since nature's legacy to them. While Upaniṣads prescribe resignation and severity of life and development of universal benevolance and love, the materialists proclaim the doctrine of uncontrolled energy, self-assertion and reckless disregard of all authority. Moral rules are conventions of men.[4]

The Cārvākas thus advocate the persuit of pleasure at any cost. They thought that although pleasure does not exist unmixed with pain, yet one should try to derive as much pleasure as possible by wisely keeping off pain and suffering. In fact they reveal their premitive, narrow look towards life. They could never grasp transcendental entity. They did not believe in any other God but a powerful king supported by the people on earth.

The Cārvākas were not only unbelievers in Vedas but they strongly criticised the Vedas in the most impolite manner. They condemned the Vedas in the roughest possible manner. At the rise of Cārvāka system, the Vedic and the Brāhmaṇic ritualism and sacrificialism had reached it peak. Too much sacrifices were made and animals were sacrified to appease Gods. In that way Cārvākas' criticism against such sacrifices was not mistaken. Some elements of rationality were certainly present in the opposition of the Cārvāka to the Vedic religion. They were responsible to a certain extent to open the eyes of the people to the facts of life. They encouraged independent and rational thinking. They were bold enough to assert their rational views against the Vedic ritualism. Whatever it may be,but they never thought of any spiritual principle 'Ātman' as such, neither they believed in the life after death. Naturally, they never believed

4. History of Indian Philosophy, Vol I : Dr. S. Rādhākrṣṇan, pp. 281-282.

in 'Mokṣa'. They reached at the another extreme point from the Vedic religious thoughts and concept of life after death.[5]

Buddhism :

Buddhism is one of the atheist schools of the Indian Philosophy. (567 B.C.—487 B.C.). 'Siddhartha', Gautam Buddha was the founder of Buddhism. Siddhartha, his own name means he who has accomplished his aim. He became Gautam Buddha. Buddha' means 'knower' or 'enlightened'. By this name he is known to his disciples and through them to the world. The four sights of the aged man bowed down by years, the sick man scorched by fever, a dead body followed by mourners weeping and tearing their hair, and the mendicant friar awakened him. He was shaken and trembled at life. Impressed by the emptiness of the things of sense, he renounced the ease, power and wealth of the palace, meditate on the eternal and open for his fellow-men an escape from the meanness of life and the illusions of the flesh. 'He made this great renunciation at the age of twenty-nine.'[6]

He renounced the world to find out the remedy for suffering. He discovered its cause and the way to stop it. He found the way to peace on earth, and preached it to the world. The theme of whole Buddhistic philosophy revolves around the four noble truths of sorrow—the existence of sorrow, the cause of sorrow, cessation of sorrow and the path leading to the cessation of sorrow.

Early Buddhism disbelieves in God as well as in Ātman (self) on various grounds, even then it is regarded as the spiritual system due to its trust in the doctrine of karma and moral perfection. It does not criticise wildely like Cārvākas on God, heaven, immortality etc. Buddha has made his statements moderately.

Early Buddhism does not believe in the existence of an

5. Ātman and Mokṣa : Dr. G.N. Joshi, p. 111.
6. Indian Philosophy Vol. I : Dr. S. Rādhākṛṣṇan, p. 348.

entity called the self or Ātman. It does not recognise any such thing as self-apart from the infinite momentary states of consciousness. It regards everything as momentary or impermanent. The world is a series of infinite momentary existences. It is a continuous series of momentary existences. Nothing in the world abides for more than one moment. Every momentary existence has all its four states, birth, the sustainance, the decay and the end, in one single moment. At every moment new existence appears and it disappears no sooner than it appears. Everyting in the world is impermanent (सर्वमनित्यम्). The world is an eternal and continuous flux of momentary existence. There is nothing permanent in this world. Permanence is only apparent and illusory, what is really existing is a continuous series of momentary existences. Buddha defined 'Reality' as that which is momentary. Whatever is real is only momentary.[7]

Early Buddhism did not accept the concept of self at all. It believed in a succession of separate and distinct stages of consciousness occurring in an orderly form. It holds that everything is anatta-not self. The self according to Buddhism is only a name given to a bundle of certain mental states like perception, thought feeling, memory, emotion, volition. Just as liquid possessing certain qualities like whiteness, certain taste and from which butter can be prepared, is given the name 'milk' so also various mental experiences are fictitionsly grouped together, and designation like 'self' is given to them.[8]

From the teachings of Buddha, his followers in their subsequent generations, drew different kinds of inferences and formed different systems by giving full scope for the free operation of their speculations. Buddhism has two main sects known as Hīnayāna and Mahāyāna. Hīnayāna means 'small or inferior vehicle' and word 'Mahāyāna' means "Great vehicle'. The ultimate goal of an adherent of Hīnayānism is the

7. Ātman and Mokṣa : G.N. Joshi, p. 117.
8. Ibid., p. 118.

attainment of his own Nirvāṇa, confirming it to oneself only, while the ultimate good of an adherent of Mahāyānism is the attainment of Nirvāṇa not only for ourself but for all beings by helping them and they were prepared to work for others even for infinite times. As compared with the Mahāyānists, the Hīnayānists required only shorter period to attain Nirvāṇa as they had confined their efforts to the individual only. They were for that purpose, regarded as belonging to lower category—The Hīnayānists. The Hīnayāna school of Buddhism represent the earlier phase of Buddhism as compared with the Mahāyāna sect, which is later development of the Buddhist philosophy. These two branches of Buddhism are still divided into sub-branches. The Hīnayāna school is divided into Vaibhāṣikas and the Sautrāntikas. both are Sarvāstivādins and believe in the reality of the external world, which we know with the help of ideas, which correspond to the external objects. The two Mahāyāna schools are Yogācāra and Mādhyamikas. The Yogācāra are also known as Vijñānavādins, who are characteristically known from denying existence of the external world and reducing it to mere ideas, and making it mind dependent. The Mādhyamika school is characterised as 'negativist' or 'nihilist' school as it proves that everying by itself is, only a void. All these schools accept the impermanance of things, the sorrowfulness of life and the necessity of attaining Nirvāṇa, as the final goal of life.

Nirvāṇa

The life of a man, teaches Buddha in his celebrated fire sermon, is a sheet of flame. Everying is on fire.[9] How? The eye is burning. The visible things, impressions formed through the eye and contacts of the eye with visible things are burning. Whether the sensations produced by the eye-contact be pleasant, impleasant or neither, they are burning with the fire of Iust , hatred and delusion, as also with the fire of birth, decay, death, grief, lamentation, suffering, dejection and despair. Similary

9. The concept of Mukti in Advaita Vedānta : A.G. Krisna Warrier, p. 75.

is the case with other sense organs like ear, the nose the tongue, the body. Eradication of all passions and lust is Nirvāna. Nirvāṇa in pali means 'blowing off' or 'to existinguish' Nirvāṇa consists in complete cessation of sorrow and it would not become unless the very cause of sorrow are uprooted. it is said in Dhammapāda "As a tree, even though it has been cut down, is firm so long as its root is safe, and grows again, thus, unless the feeders of thirst are destroyed, this pain of life will return again and again." According to Buddha to overcome and supress some desires is not sufficient but complete eradication of them is necessary. It is said "cut down the whole forest of lust not a tree only, Danger comes out of the forest of lust, when it undergrowth, then Bḥikṣus, you will be rid of the forest and free.'[10]

The seekers of Nirvāṇa are advised not to indulge in any kind of passion and desire but to weed out all the passions by means of enlightenment and proper understanding. The Nirvāṇa is described to be a state of quiescence and eternal unshakable peace. It consists in turning back upon all worldly pleasures and worldly attachments.

Buddhism believes that ignorance is the root cause of suffering. It is a kind of misconception of the world to which we foundly cling but it necessarily leads to a total disappointment and frustration. Man usually persues the right and avoids the wrong, irrational and evil. This is possible only when one acquires right knowledge of things. As man's bondage is a creation of his mind and based on knowledge, the Buddhists emphatically contend that nirvāṇa can be attained by proper knowledge which brings to an end ignorance and illusion. Ignorance consists in the absence of right knowledge of things. It is neither void nor absence of knowledge but it consists in wrong knowledge or distorted knowledge. It consists in understanding things not as they are but in understanding them otherwise than what they really are.

The Buddhists not only recognised the efficacy of

---

10. Dhammapāda MaxMuller (Tr) 338, p. 80, 283, p. 68.

philosophic knowledge but also adopted the Yogic practices as supplimentary methods for reaching the final state of Nirvāṇa. They recognised the necessity of physical postures, abstinance, tolerance, equanimous attitude, self-control, concentration and meditation as powerful means of attaining nirvāṇa. So the Buddhists regard knowledge as the real means of attaining Nirvāṇa. Nirvāṇa is intellectual enlightenment, it is enlightenment of mind, resulting from intense self-concentration and introspection, from intense abstract meditation, confirmed with the exercise of a man's own reasoning faculties and intuitions[11] Thus Nirvāṇa is a true condition of eternal rest, changeless, birthless safety is eternally attained.

Describing the nature of Nirvāṇa; Buddhists schools differ from one another. Hīnayānists think that Nirvāṇa is something existent, full of peace and beyond misery. It can be attained realising Pudgalanairātmya (i.e. realizing self or ego to be non-existent which is the real cause of pain and passion) Removing the Kleṣāvaraṇa is enough for attaining the Nirvāṇa.

Mahayānists hold that Nirvāṇa is not somethig to be achieved, it is realizing the true nature of things. It is indescribable and beyond the categories of intellect. Therefore it can not be said either existent or non-existent. Simply realising the Pudgalanairātmya is not enough for realising the Nirvāṇa but realizing Dharmanairātmya is also essential. Removing Kleṣāvaraṇa and Jñeyāvaraṇa, one can know the real nature of Nirvāṇa.[12]

Nāgārjuna being an absolutist thinks that Nirvāṇa can be said to be existent or non-existent. It is beyond all categories of thought. It cannot be defined. It is non-dual.

For *Mādhyamikas*, there is no difference between Nirvāṇa and Saṁsāra. The same 'reality' viewed from empirical standpoint is Saṁsāra and viewed from transcendental standpoint is Nirvāṇa. It is free from all pluralities. It is śūnyatā

---

11. Ātman and Mokṣa : Dr. G.N. Joshi, p. 166.
12. Mahāyānasūtrālankar of Asanga : Dr. Y.S. Shastri.

i.e. giving up all views, standpoints and predicaments. So Nirvāṇa is śūnya.

For *Vjñānavādins*, Ultimate Reality is pure conciousness. So nirvāṇa is also a state of pure consciousness, free from all subject-object duality. It is a law of universe. It is the same as Dharmkāya of Buddha. Nirvāṇa means realising this Dharmakāya. For Vijñānavādins, every being is potentially a Buddha and is in essence, the same as Dharmadhātu. It means Nirvāṇa is nothing but realization of this potentiality.

Nirvāṇa according to Lankāvatāra is something indefinable. It is non-dualistic in nature. It cannot be called even by the world 'Nirvāṇa' because it is indescribable. The real knowledge of the Ālaya is Nirvāṇa for Lankā. It is self-realization.[13]

Pure citta or consciousness is the sole 'Reality' for Asanga. It is also equated with Dharmadhātu, Tathāta and Buddhatva. Nirvāṇa, therefore, means realization of pure consciousness, which is one's own original nature. It is realization of Dharmadhātu underlying the whole universe. It is a state of omniscience which is free from all obscurations. It is a state free from all sorts of Kleṣāvaraṇa and jñeyāvaraṇa.

Nirvāṇa is purification of consciousness (which is defined by the false ideas) removing all sorts of coverings. It is discovery of the sole reality of consciousness eradicating the imaginary idea of subject-object duality, and empirical objects. Attaining Buddhahood is the ultimate aim for Asanga. It is merging with Buddhahood and becoming one with it. Becoming one with it is liberation.

Thus Asanga's Nirvāṇa is similar to Vedāntic notion of 'mokṣa' in many ways. Nirvāṇa for him is pure and all pervading. It is indescribable in terms of thought categories. This idea is similar to Vedāntic notion of 'Brahman'. Upaniṣads also proclaim Brahman as indescribable, beyond thought, beyond speech, mind and sight. Asanga says Nirvāṇa or reality

13. Ibid.

is eternal, and stable. Upaniṣads also proclaim that Brahman is eternal and stable.

Thus though Hīnayānistic concept of Nirvāṇa differs from Vedantic concept of Mokṣa, the Mahāyānists come closer to it and Asanga's concept of Nirvāṇa is nothing but Upaniṣadic concept of Mokṣa.[14]

Jīvanmukti

Buddhism recognises two kinds of Nirvāṇa (i) Sopādhiśeṣa nirvāṇa and (ii) Anupādhiśeṣa Nirvāṇa i.e. (iii) Nirvāṇa elements with the group of existence and (iv) Nirvāṇa elements without groups of existence. The Sopādhiśeṣa Nirvāṇa is Jīvanmukti and the Nirvāṇa which is attained after the destruction of groups of existence may be called as Videhamukti

In the Mādhyamika-kārika and Vṛtti also, a distinction is usually made between Upādhiśeṣa and nirupādhiśeṣa or parinirvāṇa. The description of these in the Mahāyāna Buddhhism is also the same. The Mahayānists added one more variety—the Apratiṣṭhita Nirvāṇa, the state of Bodhisattva, who shuns retiring into final release, although fully entitled to it and who by his free choise, devotes himself to the service of all beings.[15]

Dr. T.R.V. Murti while speaking of Sopādhiśeṣa Nirvāṇa says, "Of the former is the total cessation of ignorance and of the passion though the body and the mind continue to function but without passion. This state corrospond to Jīvanmukti of Vedānta and the Sāṁkhya. Buddha, after enlightenment is the representative example.[16]

Arhat (Jīvanmukta)

One who has attained Sopādhiśeṣa Nirvāṇa is called Arhat. an Arhat is one who has come over all sufferings and miseries of life. He has attained the knowledge of truth and is unattached

14. Ibid.
15. Advaitic Concept of Jīvanmukti : L.K.L. Shrivāstava, p. 89.
16. The Central Philosophy of Buddism : Dr. T.R.V. Murti. pp. 271-272.

with sense pleasure and the evils of the ceaseless ocean of life and death. But an Arhat, śrāvaka (or Pratyakṣa Buddha) aims at his own liberation and least bother for others, while Boddhisatva is ready to continue to work and to stay in Saṁsāra till the end, even for the sake of one living soul.[17]

Way to Enlightenment

According to Hīnayanists the way to get out of suffering and attain Nirvāṇa is the practice of eightfold path. To follow the eightfold way itself is the realization of Nirvāṇa. They are—right view, right thought, right speech, right action, right mode of livelyhood, right endeavour, right mindfulness, right concentration. This eightfold path is an extensive form of Buddhist trial—Śila (virtuous conduct), samādhi (concentration) and prajñā (i.e. wisdom). They also believe that 'Bodhipakṣya Dharmah' which if practised are condusive to enlightenment. It is a formula of thirty seven intellectual qualities. These dharmas are supposed to be the last percepts and injunctions of Lord Buddha.

Mahāyānists believe that a sādhaka or a monk, hankering to attain. Boddhihood, should practice six prfections (Pārmitas) during a long period of time. They are supremely pure in their nature and transcend the qualities or virtues of the śrāvaka and pratyekabuddha. They are dāna parmitas (the perfection of giving), śīla parmita (the perfection of virtuous conduct); kṣ ānti parmita (forbearance); vīrya parmita (perfection of vigour); Dhyāna pārmita (meditation); and prajñā pārmita (wisdom). This is the highest perfection attained by a Boddhisattva. It is the greatest of all pārmitā.

Buddhism recognises certain stages which a sādhaka has to cross in order to get perfect enlightenment. Both Hīnayāna and Mahāyāna agree to these stages in their own fields. Hīnayāna recognises four stages of inner satisfaction which an arhat has to cross. They are (i) Sotāpanna (ii) Sakṛdagāmi (iii) Anagāmi (iv) Arhathood. At the last stage an aspirant becomes an

17. Advaitic Concept of Jīvanmukti : L.K.L. Shrivāstava., p. 90-91.

Arhat. He becomes pure and most honourable. Mahāyānists believe that the whole career of Bodhisattva is divided into ten degrees or stages (Bhūmis). Bhūmis, they define as earth, place, region, ground, plane, stage, level, state of consciousness. The ten Bhūmis are (i) Pramuditā (joyful) (ii) Vimalā (pure) (iii) Prabhākari (light giving) (iv)Ārcismati (effulgent) (v) Sūdur-jayā (very difficult to conquer) (vi) Abhimukhi (free to face) (vii) Dūrangamā (far reaching) (viii) Acalā (steadfast) (ix) Sādhumati or Madhumati (stages of good beings of thought) and (x) Dharma Megha (Clouds).

Thus the ideal of Jīvanmukti and Mokṣa are discussed and accepted in Buddhism. A Jīvanmukta is termed as Arhat. Jīvanmukti is termed as Sopādhiśeṣanirvāṇa, arhatva, Buddhatva, Sarvajñatva etc. A Jīvanmukta is termed as Arhat, Buddha, Sarvajña, Muni, Brāhmaṇa, Vītarāga, Tattvajñāni and so on. Buddhism justifies the existence of a liberated saint by explaining further that he exists because of his compassion towards other beings; his attempt to liberate them. The essence of Bodhisatva is rightly described as 'an unequivocal affirmation of the social altruistic nature of human kind.' This ideal of Mahāyāna Buddhism is a unique contribution to Indian Philosophy. It has united in itself both spiritual and ethical experiences.[18]

JAINISM

As the Buddhists are the followers of Buddha, the awakened, the Jains are the followers of Jina, the victor, a title applied to Vardhamāna, the last prophet of Jains. It is applicable also to those men and women who have conquered their lower nature and realised the highest. The name Jainism indicates the predominently ethical character of the system.

Vardhamāna was the second son of a Kṣatriya chieftain in Magadha, the modern Behar. He was the contemporary of Buddha, was born in 599 B.c. and died in 527 B.C.

18. Ibid., p. 95-96.

Jaina tradition ascribes the origin of the system to Ṛṣ abha, who lived many centuries back. There is evidence to show that so far back as the first century B. C., there were people who were worshipping Ṛṣabhadeva, the first Tirthankara. There is no doubt, that Jainism prevailed even before Vardhamāna or Pārśvanātha. The Yajurveda mentions the names of three Tirthankaras—Ṛṣabha, Ajitnāth and Ariṣṭ-anemi. The Bhāgavata Purāṇa endorses the view that Ṛṣabha was the founder of Jainism. Whatever be the truth of it all the Jains believe that their system had previously been proclaimed through countless ages by each one of a succession of great teaches.

So Ṛṣabha was the first and Vardhamāna was the last Tirthankara. Ṛṣabha was the founder of Jainism, and the Jain scriptures are founded on the teachings of Vardhamāna Mahāvir. He was called the Jina, the spiritual conqueror, Mahāvīr, the great hero. He reformed the creed of Pārśvanāth. Jains are the followers of Jina, the victor, Vardhamān Mahāvīr.

Vardhaman's followers were drawn chiefly from the Kṣ-atriya aristocracy, and he organised them into a regular community, with lay and monistic members of both sexes.

Vedic religion had been degraded at that time. Number of animals were sacrificed to apease Gods. Serious atrocities were being committed. Neither Jains believed in God nor they accepted authority of the Vedas. They never believed in God as a creator of this world because they could explain the world without God. Another reason was that they disliked the Brāhmanic predominance which exploited God for its own supremacy.

'Ahimsā in conduct is the foundation of Jainism—Jainism stands for ahimsā in conduct, Anekānta in thought, Syādavāda in speech and aparigraha in social relations.' These are the four pillars on which the edific of Jainism stands.[20]

19. Indian Philosophy : Dr. S. Rādhākṛṣṇan Vol. I, p. 287.
20. Jain Darśan : Mahendrakumar Nyāyācārya, p. 9.

'As I do not like pain, so nobody else in the world' is the maxim which is the foundation of Jain Ethics.

Jīva (Jndividual Soul)

The whole universe is analysed into two everlasting and uncreated independent categories of 'Jīva' and 'Ajīva' (matter). The jīva is the enjoyer and ajīva is the jaḍa or the thing enjoyed. Jīva has consciousness, ajīva is devoid of consciousness. It is said in Sarvadarśanasangraha that jīva alone is conscious and ajīva has no consciousness—तत्र बोधात्मको जीवः। अबोधात्मकस्त्वजीवः।[21]

'What knows and perceivs the various objects, desires pleasure and dreads pain, acts beneficially or harmfully, and experiences the fruit thereof that is jīva. Jīva and ajīva do not correspond to I and not I. It is an objective classification of things in the universe that underlies the distinction of jīva and ajīva. Animate beings are composed of soul and body, and their souls being distinct from matter are eternal. Ajīva is divided into two main classes, those without form (arūpa) as dharma, adharma, space time and those with form (rūpa), as pudgala or matter.'[22]

Jains hold that the jīva or the soul is permanent. 'If the permanence of the soul be not admitted, someone will be the doer of actions and some other will be the reaper of their fruits.'[23] Each animate being possessed a soul or jīva. Thus there are infinit jīvas, innumerable souls, earthsouls, watersouls, air-souls, plant-souls, animal-souls, human souls, God-souls. The soul is an eternal spiritual substance. It is an immaterial principle. It transmigrates from one body to another. The ajīva is matter, the principle of rest. Matter is either an atom or aggregate of atoms. All atoms are homogenous. Karma is a kind of subtle matter which enters into the soul and causes its bondage. Matter can not be derived from soul. The soul and matter are entirely heterogenous. Thus jīva is constituted of

21. Sarvadarśanasangraha, p. 67.
22. Indian Philosophy : Dr. S. Radhakṛṣan Vol. I., p. 314.
23. Sarvadarśanasangraha, p. 48.

consciousness and it is sentient principle in each animate thing, whereas pudgala or matter or ajīva which is completely devoid of consciousness stands in direct contradiction to jīva.

jīvas are co-extensive to the bodies as lamps are co-extensive to the room in which the light is put—प्रदेशसंहार विसर्गाभ्याम प्रदीपवत्।[24] Cognition, feeling and conation are its qualities. It is the knower, enjoyer and active agent. It possesses knowledge and perception. Consciousness is not its accidental quality but it constitutes it essence. The jīva is not under control of any supra-mundane spirit. It is the master of its own destiny. It binds itself with its own predispositions and liberates itself with its own efforts.

The Jains recognise two kinds of souls i.e. the saṁsārin and the siddha jīvas. The siddha jīva is a liberated soul and the saṁsārin jīva is the soul in bondage. It is also called the mundane soul, and it is combined with Karmic matters. 'The soul in its pure state is possessed of infinit perception i.e. ananta-darśana, infinit knowledge i.e. ananta-jñāna, infinit bliss i.e. ananta-sukha and infinite power i.e. ananta-vīrya. It is all perfect. Ordinarily however, with the exception of a few released pure-souls, all other jīvas (saṁsārin) have all their purity and power covered with a thin veil of karma-matter which has been accumulating in them from beginningless time.[25] Thus the jīvas are classified into bound and liberated. The bound are again divided into (i) having mind or mobile and (ii) without mind i.e. immobile.

Bondage

Jain theory of bondage is unique in Indian Philosophy. According to them the jīvas are bound by their own karmas. With every activity mental or physical, however subtle it may be, the karmas penetrate the jīva and cover it. This is the bondage of jīva. Thus the cause of bondage of jīva is its own passionate karmas. With evil thoughts karma currents stick to

24. Dravyasamgraha II 27 & 132.
25. History of Indian Philosophy Vol. I : Desgupta, S.N., p. 189.

the jīvas and cover it. With good thoughts and actions they leave the jīvas.

According to Jainism karmas are infra-automatic particles of matter. It is subjective as well as objective. At every moment we are producing them by the activities of mind, speech and body by desire aversion and delusion. Every impression becomes karma in the time being and is sticked to the soul. Every moment, while doing actions good or evil, karma currents enter the soul. This is called āsrava in Jain Philosophy—पुण्यपापागमव्दारक्षलन आस्रवः। As water comes from the cannel, so karmas get into the soul by the cannel of yoga—यथा सरः सलिलवाहिव्दार तदास्त्रवणकारणत्वात् आस्त्रव इत्याख्यायते, तथा योग प्रणालिकया आत्मनः कर्म आस्त्रवतीति। योगः आस्त्रव इति व्यपदेश महीति।[26]

Causes of Bondage

Jainism holds that jīvas in worldly samsāri condition are in bondage and that they becomes siddhas when they attain liberation or Mokṣa. jīvas remain in bondage, so long as they are in contact with karmic matter. So āsrava is the cause of bondage. There are two kinds of bondages.

(i) Bhavabandha and (ii) Dravya-bandha. Besides these two, Jainism recognises four kinds of bondages. They are—(i) Prakṛti (ii) Sthiti (iii) Anubhāga and (iv) Pradesha-bandha.

Besides these, there are three causes of bondage, they are—(i) Perverted attitude i.e. मिथ्या दर्शन (ii) Perverted knowledge i.e. मिथ्या-ज्ञान and (iii) perverted conduct i.e. मिथ्या-चरित.

As the cause of bondage is the union of karma—matter with the soul, the libertation consists in the separation or destruction of these karma-particles. So what is necessary for the seeker of liberation is to check karma-particles which are entering into soul, which is technically called संवर in Jain Philosophy. To check āsvara is saṁvar. Saṁvar means stoppage or control—आस्रव निरोधः संवरः।

26. Rajavartikā of Akalankadeva : 1-4-16 VI 2-4.

Jains recognise eight kinds of karmas as the influx of karmas affect the soul in eight different manners. The eightfold karmas are : (i) Jñānāvaraṇīya (ii) Darśanāvaraṇīya (iii) Vedanīya (iv) Mohanīya (v) Āyu (vi) Nāma (vii) Gotra (viii) Antarya. They hold that the karmic matters enter a jīva through the senses or these may flow also through emotions i.e. kaṣāya.

Various kinds of āsravas are the real cause of bondage. The krodhāsrava, the māyā āsrava, mana-āsrava, lobha-āsrava put the soul into fetters and hinder the śoul's enjoyment of its beautiful nature. They also recognise subha and ashubha āsrava thorugh which good and evil karmas enter into the soul. If one obtains good influence by meditating upon Tirthankara, merit or puṇya flow into it. The various kinds of āsravas are nothing but those ways by which one comes to be attached to the things of worldly life, which create attachment and false impressions in him. It makes him experience pleasurable or painful experiences. Thus he moves continuously in the circle of birth and death. The original peace and tranquility is lost and thus continues to exist in bondage.

Means to Liberation

The root cause of the jīva's falling into bondage is subreption or mithyātva. The subreption is like the nescience of the Vedānta. It consists in false identification of soul and karma-matter and their attributes. So the root cause of bondage and misery is ignorance, nescience or subreption, and that one can attain deliverance from this bondage by acquiring real knowledge of the soul. So the self-realization is the path of liberation. Mokṣa can be attained by right knowledge of the soul. Thus a right understanding of the soul alone can deliver one from the bondage of the jīva. No karma can bring to an end karma. On the contrary cessation of the operation of karma frees the soul forever from the fetters which cause to him afflictions.[27]

---

27. Ātman and Mokṣa : Dr. G.N. Joshi, p. 245.

There are six means to stop the influx of Kārmic (saṁvara) namely (i) Gupti (ii) Samiti (iii) Dharma (iv) Anuprekṣa (v) Prisahajaya (vi) Charitta. It even destroys the karma-matter which has already entered into the soul, before the practice of penance. Nirjarā is necessary for the attainment of mokṣa. Nirjarā means throwing off the karma-matter from the soul. It means like complete nutralization of karma-matter. Nirjarā are of two kinds (i) Avipāka nirjarā and (ii) Savipāka nirjarā. Completion of the process of nirjarā leads to Mokṣa. After the absolute separation of karma from the soul, the soul is liberated. Completion of the process of nirjarā is just logically prior to Mokṣa. Factually both happen at the same time. After the complete separation of the karmas from the soul three things happen simulaneously—

(i) Separation of the soul from body.
(ii) Upward flight of the soul.
(iii) Reaching to the summit of the Mundane space.

Thus, as soon as the cause i.e. karma particles leave the soul, the soul leaves the body, and becomes liberated (mukta).

Jainism preaches for a seeker of liberation the necessities of practising positive virtues like forgiveness to all without feeling the least anger towards anybody for any offence or humility and modesty towards all by controlling one's arrogance and pride. It preaches simplicity, honesty, the noble path of overcoming evil with good, with pure means, winning the opponent by love and foregiveness, or by changing heart. The seaker of liberation must be completely free from greed.

Liberation

Complete deliverance from veil and covering of karma is called Mokṣa or nirvāṇa. Cessation of operation of karma frees soul forever from fetters, which cause him affections. (सर्वावरण विमुक्ति मुक्ति). Right understanding of the soul alone can deliver one from bondage of jīva. No karma can bring to an end karma. According to Jainism the jīva is constitutionally free and potentially divine. The jīva in the state of nirvāṇa enjoys a state of pure and infinite knowledge and infinite perception i.e. ananta-jñāna and anant-darśana. The liberated souls are above the Gods. They are never born again. They have no longer any connection with the world and exert no influence on it.

Nirvāṇa or deliverance is not annihilation of the soul, but its entry into a blessedness that has no end. It is an escape from the body, though not from existence. The liberated is not long, nor small---nor black, nor blue, not bitter, nor pungent, neither cold nor hot-----without body, without rebirth---he perceives, he knows, but there is no analogy, its essence is without form, there is no condition of the conditioned.[28]

According to Jainism, any person who accepts jain creed can turn a wondering medicant at any stage of his life and that Mokṣa can be attained by all persons irrespective of their cast provided that they lead that type of life which is prescribed by Jainism.

'The intrinsic nature of soul is to move upward. As the nature of piece of wood is to remain on the surface of water, and the nature of flame is to burn straight way upward. As wooden things while plastered with dust remain under water, but as soon as the plaster is dissolved, comes on the surface of water, or as the flame of a lamp in ordinary course burns upwards, similarly, the soul when separated from karma particles returns to its intrinsic nature and moves upwards.'[29]

The liberated reaches to the summit of the Mundane space, which is technically called 'siddhaśītā', the realm of free and liberated. It resides there forever. It cannot move higher than that for there is no principle of motion there. From vyāvahārika point of view, such a siddha has a shadowy form which is a bit small in size than its immediately previous body.[30] He is the embodiment of Samyak-Darśana, Samyak-jñāna and samyak-carita. Thus the siddha is an embodiment of the tri-ratna and has got the perfection corrosponding to the karmas and resides

28. Indian Philosophy Vol. 1 : Dr. S. Radhākṛṣṇan, p. 332.
29. Rajvārtikā X 7.6.
30. Ibid., V. 16-8-9.

at the summit of the Mundane space. These three jewels constitute the pathway to Mokṣa.

Like other moral categories, Jain sages classify Mokṣa into (i) Bhava-Mokṣa and (ii) Dravya-Mokṣa. When the soul becomes free from the four ghātiya karmas (destructive karmas), it is said to have attained subjective liberation (Bhāva-mokṣa) and when all the non-destructive karmas are also destroyed, it attains objective liberation (Dravya-mokṣa).

According to Jain, there is a ladder of fourteen steps by which a jīva has to pass through the stages to reach 'Mokṣa'. These stages of spiritual development are called guṇasthānas in Jain philosophy. They are—

(i) Mithyāḍṛsti guṇasthāna
(ii) Granthibheda
(iii) Miśra Guṇashāna
(iv) Avirata Samyak Dṛṣti
(v) Desavrjati guṇasthāna
(vi) Pramatta Sāmyatā
(vii) Apramatta Sāmyatā
(viii) Apūrvakaraṇa. or Nivṛtt
(ix) Anivṛtti karaṇa
(x) Sūkṣma Samparāya
(xi) Uṣatamoha
(xii) Kṣīṇakaṣāya
(xiii) Sayoga Kevalī
(xiv) Ayoga Kevalī

These are the fourteen stages which a soul has to cross before final liberation.

In the state of Nirvāṇa the soul realizes the ananta-chatuṣ-ṭaya, which is its intrinsic nature. They are infinite knowledge (ananta-jñāna), infinit faith or perception (ananta-darśana), infinite bliss (ananta-sukha) and infinite power (ananta-vīrya). The Jains are divided into two sects (i) Śvetāmbara and Digambara. According to Digambaṛa, women cannot attain 'Mokṣa'.

One who has got keval-jñāna (omniscience) is kavalī (perfect). 'kevalī' in Jainism is Jīvanmukta, who has destroyed the karmas that obscure the inherent capacities of the soul and has freed himself from bhāvabandha. Tirthankaras who are worshipped by Jains have bodies even after attaining omniscience or kevalajñāna. 'Vītarāga', 'Jina' and 'Arhat' are also used as Synonyms of a Jīvanmukta. The word 'Jīvanmukti' itself has been used in 'Prameya-kamala-Martanḍa' and 'Nyāya-kumuda-chandra' of Prabhāchandra (between 980 A.D. to 1065 A.D.) to denote the state of a kevali.[31] So Jainism accepts Jīvanmukti.

Jainism raised voice against cast hirarchy and made the way to turn ascetic easier, more practicable and acceptable to the persons of all casts.

Nyāya-Vaiśeṣika

The Nyāya and the Vaiśeṣika systems of Indian Philosophy are theistic systems. They believe in the Vedas but at the same time give a new cosmology. These two systems are originated by two different philosophers namely Nyāya by Gotama and Vaiśeṣika by Kaṇāda. Nyāya is mainly busy with the problems of logic and epistimology whereas Vaiśeṣika is busy with the problems of ontology and metaphysics. Nyāya accepts the Vaiśeṣika metaphysics and Vaiśeṣika accepts the Nyāya epistemology and logic. So these two systems are thought together and are taken similar.

The Nyāya philosophy has been held in great reverence for a long time past. Manu includes it under śruti. Yājñāvalkya regards it as one of the four limbs of the Vedas.[32] The classical study of the Hindu comprise the five subjects of Kāvya (literature), Nātya (drāmā), Alankāra (rhetoric), Tarka (logic) and Vyākaraṇa (grammer). Whatever other specialised studies a student may take up later, the primary course includes logic, which is the basis of all studies. Every system of Hindu thought accepts the

31. Advaitic Conept of Jīvanmukti : L.K.L. Shrivāstava, p. 80.
32. Yājñāvalkya smriti : 1.3.

fundamental principles of the Nyāya logic and even in criticising the Nyāya system, uses the Nyāya terminology and logic. The Nyāya serves as an introduction to all systematic philosophy.

It is a philosophy of distinctions. Its standpoint is scientific and analytic. 'The Vaiśeṣika system takes its name from 'viśeṣa' which means Difference. It insists that it is in the perticular imperceptible souls and atoms that true individuality is to be found. The Vaiśeṣika is essentially a philosophy of distinctions since it does not tolerate any attempt at dissipating the independence of self and objects in a supposed more perfect individuality. Its standpoint is more scientific than speculative, more analytic than synthetic, though it is not able to set aside questions about the general character of the universe as a whole. The Vaiśeṣika and the Nyāya agree in their essential princples such as the nature and the qualities of the self and atomic theory of the universe, yet the classifictions and characterization of the categories and the development of the atomic theory give to the Vaiśeṣika a distinctive interest and volume.'[33]

The approach of the Nyāya-Vaiśeṣika to the philosophical problems is thoroughly rational. The acceptance of the authority of Śruti as pramāṇa is merely a leap homage probably only to save themselves to be ranked with Nāstiks. The rational approach is emphasized so much that they have forgotten that there can be certain problems which reason cannot solve. Even the self has not been taken for granted. They have establshed the existence of self by the help of logic. 'I' in the expression ' I am happy' refers to self. It cannot refer to any material object like body, mind etc., because they are all unconscious and the category which is the ground of our experience should be conscious.[34]

Thus the Nyāya and Vaiśeṣika systems proves the existence

---

33. Indian Philosophy : Dr. S.Radhakriṣṇan Vol. I, pp.166-167.
34. Comp. Study of the Concept of Liberation : A.K. Lad, p. 86.

of self with the help of logic. According to them the soul is eternal. The reason for that is—even a newborn infant experiences joy, pleaure, pain etc. which is possible only because of the continuity of memory of what has been experienced earlier. This proves that the soul is eternal. Desire, aversion, effort, pleasure pain and cognition are special qualities of the soul. The soul cannot be identified with the body, sense organs etc, because they are all perishable, but the soul is immutable. So it is something distinct from the body, sense organs, mind etc. According to Kaṇāda also, all activities of living body indicate soul. They all cease to function when the soul is dissociated with the body.

The approach of Nyāya and Vaiśeṣikas being rational, realistic, they believe in the plurality of souls. They hold that there are as many souls as there are bodies. The theory of one soul has been rejected by them on the ground of 'existing differences in the universe'. In the world some souls are in bondage, while others are liberated. Had there been only one soul, with the liberation of one all others have been liberated. This proves that there are many souls, as many souls as are the bodies. Similarly there are differences in the world between rich and poor, happy and unhappy. If there is only one soul, no satisfactory explanation of these existing differences can be given. Again in order to reap the fruits of good and bad actions the soul has to pass from one body to another. The cycle of birth and death implies that the soul has permanent eternal existence. The place of enjoyment of the soul is body. 'The soul is the resort of the experience like pleasure.' (सुखाद्याश्रयत्वं जीव लक्षणं।).[35] The endurance or persistence of the soul can be proved also on the moral basis. 'If the body were soul, there should be release from sins as soon as the body were burnt.' (शरीरदाहे पातकाभावात्।).[36] Again according to Nyāya-Vaiśeṣika systems, there are certain immaterial and incorporeal things

35. Tarkasangraha : Annambhatta, p. 17.
36. Nyāyasūtra of Gotama (Tr.) 4, p. 63 : Vidyābhuṣaṇa S.C.

such as knowledge, desire, will, pleasure, pain etc. They are passing phases of mental expriences, having momentary existences. These are the attributes of 'soul'. 'Desire, aversion volition, pleasure,pain and intelligence are the masks of 'soul'.[37]

(इच्छाव्द्वेषप्रयत्न-सुखदुःख-जानामि आत्मनो लिङ्गमिति।).

As colour belong to fire, taste to water, touch to air, odour to earth and soul to Ākāśa, as the attributes of each substance belong separately, so do these attributes like knowledge, desire, hatred, pleasure and pain belong to a separate substance called soul. The soul is regarded as a substance which acts as a possesser of the mental attributes. Kaṇāda recognises these qualities as the sign of soul.[38]

Though the soul inhers all these qualities, it does not possess these attributes as its permanent qualities. Even consciousness is not its essential and permanent attribute. All the physical attributes arise in the soul when it comes into contact with mind and body. In itself it is not constituted of eternal consciosness as Advaita holds. Consciousness is its quality which sometimes arise in it and sometimes absent from it. Consciousness is produced in the soul when it gets connected with the body thorugh mind. The soul has simply the potency of knowledge.

Nyāya-Vaiśeṣika recognise nine final and eternal realities which are mutually irreducible and explain the finite varieties of the universe by their properties and relations. The nine elements (dravyas) are (i) earth (ii) water (iii) fire (iv) air (v) space (vi) time (vii) direction (viii) the self (ix) mind. All these substances are not material but each of them has quite a distinct characteristic which is not possessed by any other substance. The dravyas exist in the form of their infinitesimal parts, known as the aṇus or atoms. Both these systems believe that the ultimate reality consists of the minute particles or atoms. The concrete things of the world are infinitly divisible. On a thorough

37. Ibid., 1.1.10.
38. Vaiśeṣika-darśana : Kaṇāda sūtras (p.7) : Praśastapāda (com.).

going analysis of the things of the world, we arrive at the same existence which becomes final and which remains further indivisible. Thus according to these schools, these final indivisible and indestructible things to which we can reduce the gross and concrete things of the world are the atoms. They form the points of termination. They are uncaused things, and act as the constituent causes of all the material things of the universe, having different magnitudes. They are simple, partless, and are perceptible to our sense-organs, extermely subtle. They are to be understood by inference and imagination as they can not be the object of our perception or direct knowledge. They are the fundamental metaphysical points with their special peculiarities.[39]

'The doctrine of 'substance' put forth by this system seems to be a reaction to the Buddhist universal impermanance of things. The Buddhists had reduced the whole world to a continuous series of momentary existences, and had denied the existence of permanent substance behind the momentary qualities. The Nyāya-Vaiśeṣika theory of substance appears to be an authentical and powerful reaction to the Buddhist impermanance.[40]

Two Kinds of Ātman

These systems believe in the two kinds of Ātman. One is supreme self and the other is individual self. i.e. Paramātman and Jīvātaman. The Supreme Self is not only omniscient, all pervading, eternal but his knowledge also is all-pervading, whereas individual soul also is all pervading, eternal but each jīvātman is separate from each other. As many bodies are there, so many souls are there. This soul is the real doer of action. It is said in Bhāṣā Parichcheda about the soul that—'The soul is the inspirer of organs etc. for an instrument requires an agent.'[41]

Naiyāyikas hold that the soul reveals itself to us in the form of ego, 'I hood.' (अहंप्रत्यय).

---

39. Ātman and Mokṣa : Dr. G.N. Joshi (p.268).
40. A History of Indian Philosophy : Vol.-1 : Dr. S.N. Dasgupta, p. 274.
41. Bhāṣāparichcheda : Swami Mādhavananda Sūtras 47., p. 65.

Bondage and its Cause

The Nyāya system is known for its science of correct thinking, science of tarka or reasoned arguments. It's logic is the pre-requisit of all philosophical thinking. This philosophical thinking leads to knowledge. The purpose of knowledge is attainment of liberation from all pains and sufferings of life. The Naiyāyikas hold that unless the nature of pain and its root cause are properly apprehended, it is not possible to free ourselves from the pain and sufferings of life forever. The aim of knowing is the attainment of knowledge and the aim of knowledge is attainment of liberation. The soul, the body, the senses, objects of senses, intellect, mind, activity, pain, release, transmigration are the objects of knowledge.

Misapprehension of these objects cause bondage and its consequent suffering in human life. The Naiyāyikas lay emphasis first upon the need of properly grasping the logical interrelation of these objects and then by going to the root cause of pain. These systems do not lay stress on performing rites, rituals, nor on idol worship. They emphasize on pure understanding of the 'Reality'. Like Advaita-Vedānta, they believe that liberation can be attained by right knowledge and not by other methods.

Release which consists in the souls, getting rid of the world which is the condition of Supreme felicity marked by perfect tranquility and not tainted by any defilment. A person by the true knowledge of the sixteen categories is able to remove his misapprehensions. When this is done, his faults such as affections, aversion, stupidity disappear. He is then no longer subject to any activity, and is consequently freed from transmigration and pain. This is the way in which his release is effected and Supreme felicity secured.[42]

Nyāya system agrees with the Vedānta who believe that the ignorance or the misapprehension is the root cause of all evils in the world. Vātsyāyana in his commentary on the sūtras

42. Ibid., Sūtras 1-2.

of Gotama describes misapprehension or mithyājñāna is not simply as the absence of right philosophical knowledge, but something positive, something like the infatuation out of which the world is generated. He writes, "The infatuations of the sense of 'I' or ego is to feel the sense 'I hood' in objects which are devoid of self 'Egoism' (i.e. ahankāra) consists in seeing in the selfless or material things our ego."[43]

Material objects and immaterial souls are two entirely different entitites from each other. They are categorically different things. But we confuse our soul with the non-spiritual objects. We include our body or mind in the ideas of our soul. Naturally we identify ourselves with them, as they are liable to frequent disappointments and sufferings, we also suffer. In fact suffering cannot belong to innermost self.

The real cause of production of body lies in the effects of the acts of the past life of an individual. As are the acts, so are the natures of our begotten bodies. The whole cycle of birth and death and the continuity of it with the successive birth is guided and regulated by the doctrine of karma or by the desert of an individual. So this karma-matter is responsible for maintaining the consistency of relation between deeds and their effect. 'The body is produced as the fruits of our previous deeds.'[44]

Nyāya-Vaiśeṣika hold that a man who seeks pleasure in pure form. Pleasure, howsoever pure it may be, to some extent it is mixed with pain. The two are inseparable. Such a pleasure cannot bring to a man the everlasting happiness and satisfaction. "Pain the cause of uneasiness is the sign that the soul is not at rest with itself. The highest good is deliverence from pain and the enjoyment of pleasure for the pleasure is always mixed with pain." Hatred of pain is still hatred and the attachment of pleasure is still attachment, and so long as these are operative, the highest good is beyond reach."[45]

43. Nyāya Sutra by Gotama : Vātsyāyana (com.).
44. The Nyāyasūtra of Gotama : Vidyābhṣaṇa S.C. Sutra 3.2.64.
45. Indian Philosophy (Vol. II) ppp.161, 162, 163.

Thus the human beings live. They live means they act. To live means to act some way. These activities produce good and bad effects which cause another birth. This continuous cycle of birth activities, acquisition of karma, death, next birth and death form the ever revolving wheel of saṁsāra. Liberation frees human beings from such everlasting wheel of life. Liberation means complete freedom from pain and suffering. "The state of release is absolutely free from distress as that of dreamless sleep. As there is no distress in a person who is sound asleep and sees no dream, so there is no association of troubles in one who attains release."[46]

Mokṣa—(Apavarga—Niśreyas)—Mokṣa is known as "Apavarga' in Nyāya and "Niśreyas' in the Vaiśeṣika system. It is said in Sarvadarśanasangraha that—'Liberation is attained by philosophical knowledge or indeed liberation occurs because of philosophical thinkings."[47] The motive and end of philosophical understanding of the world is liberation. The purpose of the study of the right understanding of the world and life is the attainment of liberation.

Mokṣa or Aparvarga means complete cessation of specific qualities of the ātman. Complete cessation of pain and sorrow is said to be the characteristic of liberated ātman. In "Apavarga' i.e. in Nyāya liberation, there is utter absence of pain and suffering and so also that of happiness and pleasure. They can not abstract pleasure from pain. They believe that pure pleasure devoid of pain is not possible, and hence cannot imagine that in liberation pleasure or happiness of any sort exist. Liberation means painlessness and pleasurelessness also. 'Pleasure being associated with pain in equivalent to pain, their association means their inseparability. Just as honey mixed with poison becomes poison, so becomes pleasure, pain mixed with the pleasure.'[48]

---

46. The Nyāyasūtra of Gotama : Vidyābhūṣaṇa sutras 4.1.63.
47. Sarvadarśanasamgraha, p. 246.
48. Tarka-saṃgraha : Keśava Miśrā, p. 92.

So liberation is a state of neither pleasure nor pain. It is a state of indifference and neutrality. The state of liberation according to Nyāya-Vaiśeṣika lies in proper apprehension of the nature of "soul' and "reality'. Mental ties or bondage can be entirely broken by the enlightment. The pain of liberation passes through the philosophical understanding of the self and the world. Vātsyāyana describes in his commentary the stages of the attainment of liberation and shows the causal chain of various factors that lead to the final liberation. He writes, 'when the misapprehension disappears, because of its disappearance faults also disappear. With the disappearance of faults disappear the tendencies to activities. With the disappearance of tendencies disappear birth and with the disappearance of birth disappear pain and in the disappearance of pain lies absolute liberation, or final beatitude i.e. niśreyasa.[49]

Knowledge of Reality and performance of duties both are necessary for liberation. Liberation is a combination of knowledge and duties. The prohibited actions and prudential duties should be discarded. But compulsory duties and the occasional duties should be performed. These duties purify the mind. Non-performance of them will produce the sin of omission which will require another birth to be exhausted.

Like other schools of Indian philosophy, the value of yogic practices is recognised as helpful in the attainment of liberation. Past, present and future karmas are exhausted forever by practicing the yoga.

Jīvanmukti

Nyāya-Vaiśeṣika do not believe in the embodied release. According to them the only cause of bondage is embodiment, so the conception of Jīvanmukti is a contradiction in terms as it is for Rāmānuja and other Vaiṣṇava-Vedāntins.

Saṁkhya-Yoga

The Saṁkhya system is one of the most ancient system in

---

49. Ātman and Mokṣa : Dr. G.N. Joshi, p. 313.

the history of ancient philosophy. This system advocates the ontological dualism of Prakṛti and Puruṣa. The dualism of Puruṣa and Prakṛti is the fundamental doctrine of the sāṁkhya system. Kapil Muni is said to be the first originator of this system. Sāṁkhya derives its name from the word 'saṁkhyā'. The 'Saṁkhyā' is used in the sense of thinking, knowledge, counting etc. or the word saṁkhyā means 'perfect knowledge'. The system is called sāṁkhya since it gives perfect knowledge of the self, as quite distinct from 'Prakṛti' and its evolutes, body, sense-organs, mind, intellect and egoism which annihilates all kinds of sufferings.[50]

Pandit Gangānāth Jhā is of the opinion that, 'sāṁkhya means knowledge of self through right discrimination.' Guṇaratna thinks that Sāṁkhya derives its name from its first founder 'Saṁkhā'.[51]

Sāṁkhya system is as old as the Buddhistic philosophy. It also bears resemblence with the Upaniṣadic Philosophy. As it contains some elements of Upaniṣads, it is probable that it came into existence in the post Upaniṣadic period.

Sāṁkhya system is atheistic in character. They do not refuse the existence of God, for the sake of refusing it, but they refuse its existence for not having sufficient ground to accept him. The Sāṁkhya atheism has its own peculiarities. The existence of God is denied not because God does not exist, but because of the difficulty of proving God's existence.[52] Sāṁkhya did not believe absolutely in the authority of the Vedas. It did not support the Vedic ritualism and sacrifices. It did not believe that these vedic rituals and sacrifices are necessary for the attainment of liberation. It laid more stress on the knowledge of 'Reality' and realization of spirit than any kind of ritualsm. It had its own independent approach to 'Reality'. Kapil had taken quite independent stand in his enquiry of truth. He

---

50. Outlines of India Philosophy : Jadunāth Simha., p. 254.
51. Introduction to Tattvakaumudi : Gangānath Jha, p. 2, 3.
52. The six systems of Indian Philosophy : Max Muller, p. 365.

interpreted the world in his own way. His approach towards the problem is novel and original.

The approach of the Sāṁkhya-Yoga system to the philosophical problem is purely rational. The acceptane of authority of the Śrutī is only a lip homage 'In Kapil's doctrine, for the first time in the history of the world the complete independence and freedom of human mind, its full confidence in its own powers were exhibit.[53]

Like other systems of Indian Philosophy prticularly Buddhism, Sāṁkhya admit the universal sufferings. Our ultimate goal is to be free from all sorts of sufferings. Yoga also admit that the worldly life is full of sufferings. It emphatically maintains that mere philosophy cannot save us, neither the subtletics of things are enough, what is more necessary and important thing is the 'control of will'.

The Sākhya Concept of Puruṣa and Prakṛti

The Sāṁkhya philosophy believes in the absolute dualism of Puruṣa and Prakṛti i.e. the spirit and matter. They are exclusive of each other, and collectively exhaust the sphere of being. There cannot be anything in this world which is not either of these two. The spirit or soul or Puruṣa acts as the source, substratum and ultimate resort of all the things in the world. Except consciousness, everything that exists in the world arises out of Prakṛti, the original substance. It subsists in Prakṛti and ultimately after destruction returns to Prakṛti. Prakṛti or nature acts as the material and efficient cause of the world. It is all pervading. Whatever is in the world, is the modification of the original primordial matter or Prakṛti.

Sāṁkhya maintain the plurality of Puruṣas on the ground that we experience this in our everyday life, but this pluralism is only quantitative, qualitatively all Puruṣas are alike. Had there been only one Puruṣa, the birth and death of one would have meant the birth and death of all. Similarly, the bondage and

---

53. Philosophy of Ancient India : Dr. S. Radhakriṣṇan Vol. II, p. 249.

libration of one, would have bound and liberated all at the same time. But we do not experience such. So it is established that there are many puruṣas. Consciousness is the esssence and not the adventitious quality of Puruṣa as it is for the Ātman of Nyāya-Vaiśeṣika.

Bondage

According to Sāṁkhya-Yoga, bondage is caused due to ignorance, wrong knowledge. The ignorance consists in not knowing the true nature of one's own self. It is because one fails to discriminate between Puruṣa and Prakṛti. It is to confuse Puruṣa for either Prakṛti or for any of her evolutes. In fact Puruṣa is never in bondage. It is ever free. It is said in Kārika. "Verily not any soul is bound, nor is liberated, nor migrates, but nature alone in relation to various beings is bound and released and migrates."[54] According to Sāṁkhya bondage is of three kinds (i) Natural (ii) Evolutional and (iii) Personal. when Puruṣa knows its own nature that it is not Prakṛti but pure consciousness, it gets free from all bondages.

Mokṣa—Kaivalya

Kaivalya means isolation. Mokṣa is a state where Puruṣa is absolutely isolated from Prakṛti and her evolutes. Like other systems of Indian Philosophy perticularly Buddhism, Sāṁkhya admits the universal sufferings. Our ultimate goal is to be free from all sufferings. Kapil does not enter into a minute analysis of the conception of liberation. Like Buddha, the object of Kapil was to show the origin and destruction of the worldly pain. But the followers of Kapil developed the conception of liberation. The cause of bondage is that Puruṣa wrongly identifies itself with Prakṛti and her evolutes. When this wrong identification ceases by right knowledge, it is perfectly free from the clutches of Prakṛti.

The state of liberation is also known as 'apavarga' in Sāṁkhya system. "The self in this state not only has no pain or

54. Sāṁkhya Kārikā : Iśvarakṛṣṇa : Com. by Gauḍapāda (Tr. H.T. Celebrooke) 62/235.

pleasure in that condition, it is also without knowledge for it has not the means namely buddhi and its accessories wherewith to know.[55]

The soul can never be in bondage because by nature it is free from all attachments. Out of non-discrimination the soul appears to be connected with the Prakṛti but in fact Prakṛti cannot bind it. When Prakṛti is active, it catches the reflection of Puruṣa and casts its shadow on the Puruṣa. The Puruṣa is active only by the reflection of Prakṛti in it. Puruṣas are not actually affected by the various plcausrable and painful experiences of the Prakṛti. When the Puruṣa detaches itself from the Prakṛti, it resides in the loneliness without standing in the need of any other thing besides itself. It remains thoroughly content in itself. When Prakṛti ceases to act, the modifications of buddhi cease and the Puruṣa enjoys its original nature. The soul or Puruṣa remains lost in itself. This is apavarga or Kaivalya or mokṣa for Sāṁkhya system.

Jīvanmukti

We get many references of 'Jīvanmukti' in the authoritative works of Sāṁkhya and Yoga system. The earliest possible work on the Sāṁkhya is the 'Sāṁkhya-Kārikā' of Īśvara-kṛṣ ṇa (3rd century A.D.). Here we don't find the word 'Jīvanmukti' but the state attuned to it is definitly described. The commentary on Sāṁkhya-Kārikā by Vācaspati Misra namely Tattva-Kaumudi is regarded as the most authoritative work which gives clear account of Jīvanmukti. Here also the word jīvanmukti is not used. 'Sāṁkhya-sūtra' uses the word 'Jīvanmukti' and gives more vivid picture of it. So also Patañjali's Yoga-sūtra.[56]

Yoga Philosophy of Patañjali holds a unique position in Indian philosophy, because of its yogic practices. Yogic practices which Patañjali has advocated consists in cultivating physical and mental discipline .It is external and internal both. Human mind is never steady. It is often distracted and disturbed by the

55. Ātman and Mokṣa : Dr. G.N. Joshi, p. 367.
56. Advaitic concept of Jīvanmukti : L.K.L. Shrivāstava, p. 74.

various mental states. So the first necessity is for making one's mind quiet and to understand such various mental states and to control them properly. It is with the help of aṣṭangayoga that one can control his mind. With the pratice of aṣṭangayoga one possesses a genuine contentment. The practice of astanga-yoga, the knowledge of "Reality' enables him to become a Jīvanmukta. The state of jīvanmukti is characterised by cessation of all coverings of affliction and action. It is a state before the attainment of absolute isolation. This is jīvanmukti.

Pūrva Mimāmsā

The Pūrva Mimāmsā system is the most prominent orthodox system among all the Indian systems. It believes in the infalliability of the Vedas. It tries to understand properly the thoughts embodied in the words of the Vedas. The Mimāmsā regards that the vedas are eternal, self-created. It is called as 'Purva' Mimāmsā, as it deals with the former part such as karma-kānda, rituals, sacrifices of the Vedas. The 'Uttar Mimāṁsa' is 'Uttar' as it deals with the latter part of the Vedas such as jñānakanda etc. Pūrva-Mimāmsā believes that a disciple is first competent for action only and then for knowledge.

Jaimini is regarded as the earliest Mimāmsā-Philosopher. His sūtras were interpreted by the scholars like Śabarswāmī, Prabhākara, Kumāril Bhatta, and the works of these commentators were further critically presented by Parthasārathi Miśrā, Manḍana Miśrā etc. Śabarabhāṣya explained by Prabhākara and Kumārila differ in their interpretation and form two schools of Mimāmsā namely Prabhākara school and Kumārila school.

God

Mimāmsā system gave so much importance to sound and words which are formed in the Vedas that they even denied the existence of God. This system regards that Vedic texts are etrnal, hence God is not necessary. It didnt believe in God as creator, presever, destroyer of the universe. This idea that the God is the creator of the universe can not be proved and hence rejected by Kumārila. In this sense, this system is atheistic one but in the different sense, the system is polytheistic. The Mimāṁsā deals with the rituals and sacrifices primarily. Offerings are made to different deities in sacrifices. So there are many deities in whom a Mimāmsā has to put faith.

Self

The necessity of the existence of self is quite obvious. They believe in performing sacrifices. Sacrifices are performed by a performer and in most cases, they are performed to acquire some advantage in the other world. It clearly shows that they believe in the eternal entity 'self'. It does imply that this eternal entity is apart from senses, body, mind which will enjoy the fruits of the actions in future. If there is no such eternal entity, then the Vedas have made an absurd statement, but Vedas never make any absurd statement. Thus Pūrva-Mimāmsā system infers the existence of self from the scrifices that are performed. As Kumarila Bhaṭṭa says, "Vedas have declared that the results of sacrifices pertain to the performer in some birth or the other and if the soul were nothing more than mere idea, then it could not have the character of the performer (of actions) and enjoyer of results. Moreover he says, "Therefore it is with a view to establish the authority of the Vedas, that the existence of soul is to be proved here."[57] Śabaraswāmi accepts the reality of the permanent cogniser which is 'known by itself and incapable of being shown or shown by others.'

The Pūrva-Mimāmsā system holds that there is some such entity as soul that is different from senses, body, mind cognition and that it goes to heaven by the performance of rituals prescribed. The soul is the agent of actions and reaper of their fruits. It is eternal, without any change, all pervasive immaterial and omnipresent. Knowledge is not its essential nature but a mode of it. The soul is abode of knowledge, intelligence and experiences like pleausre, pain etc. These cannot be the characteristics of body.

57. Ślokavārtikā : Kumarila Bhaṭṭa, Gangānāth Jha Section 18/p.383.

Plurality of Souls

Souls are many in number. Each person possesses his own soul separately to maintain his independence. If the souls were not many, there would have been confusion of works and their rewards. Kumārila Bhatta and Prabhākara both admit plurality of souls. If there were only one soul for all persons, all would have been happy with the happiness of one and all would have been suffered with the suffering of one person. All would have been liberated with the liberation of one. On the contrary, we find in actuality that every person has his own separate experience of pleasure, pain, love, aversion, knowedge and ignorance. It reveals from the diversity of such experiences that there are many and separate souls one for each body. The Purva-Mimāmsā system believes in the doctrine of karma or adṛaṣta, it is therefore logically driven to accept the plurality of soul.[58]

The Soul is the Doer of Action

The soul is the doer of action and enjoyer of their fruits. In order to maintain the consistency of the doctrine of Karma, the doer of action must be the same to experience the consequences of action done by him. They cannot be two different persons. The soul has to be performer of sacrifices because it has to go to heaven. The old Mimāmsakās say that it cannot rise to heaven unless it performs the necessary sacrifices prescribed. According to the new Mimāmsakas the end of performing sacrifices is the attainment of 'Mokṣa' and not reaching the heaven. But the person who attains Mokṣa must be that person who undertakes the efforts for it. The doer and the doer alone can be the enjoyer. The agent is generally supposed to be involved in movement due to his activity. But Kumārila Bhatta denies any movement in the agent himself. He proves it with the help of illustration—Just as Deodatta, though he cuts many other things with his sword, it does not follow that he himself is cut by it. Similarly the soul may act as an agent to other thing by applying movements, but from that it

58. Parthasārathi Miśrā : Śāstradīpikā : 1.11.5, p. 125.

does not follow that it itself undergo any change. The soul is therefore unmoved mover of things.

Thus due to good and evil deeds the soul acquires merits and demerits which ultimately bear their corresponding results in accordance with which the soul takes birth in the successive lives. When total stock of the accumulated karma is completely exhausted, when the soul acquires no new karma by deeds, the soul becomes liberated from the bond of Saṁsāra. The soul becomes 'mukta'. Then the liberated soul enjoys this natural state forever.

Liberation : Apavarga

Jaimini and Śabara Swāmi did not face the problem of ultimate release. They pointed the way to life in heaven, but not the freedom from Saṁsāra. They were more concerned with finding the real meaning of the mantras of sacrifices. Dr. Rādhākrishnan points out that the later Mimāmasakas like Prabhākara and Kumārila Bhatta dealt with the problem of Mokṣa as the thinkers of the other systems were thinking over it seriously. The theories of liberation, therefore, are the contributions especially of Prabhākaras and Kumārilas.

Even Prabhākara was mainly concerned with ritualistic aspect but his followers gave a clear idea of Mokṣa. According to them, liberation or mokṣa is the removal of merits and demerits, which results from action. When the Dharma and Adharma cease to exist, the cessation of existence leads to the cessation of rebirth. Therefore liberation is described as 'the absolute cessation of body caused by the disappearance of Dharma and Adharma. The Prabhākaras, thus, hold that once the total accumulated stock of the desert is exhausted either by enjoying it or is destroyed by knowledge, the soul does not take fresh birth and thus gets permanent freedom from Saṁsāra. Prathasārathi Miśrā also defines 'Mokṣa' as withdrawal from Saṁsāra.[59] In liberation soul ceases to have any connection with the body.

59. Śāstradīpikā : Parthasārathi Miśra : 1.1.5, p. 130.

The Process of Attaining Liberation

The worldly life is supposed to be full of pain and sufferings. Man frequently has to undergo tremendous sufferings in the worldly life. Even the plesures that he enjoys are found to be mixed with pain. No pure or unmixed pleasure can ever be experienced in the worldly life. One gradually becomes desperate and beings to loose interest in such pleasures. A man always seeks for an agreeable experience. He moves in search of such a place where either the pleasure is not accompanied by pain or at least pain has come to an end. He finds such a hope of terminating his miseries or sorrows in Mokṣa or liberation. When he properly grasps the nature of Mokṣa, he stops performing prohibited actions. Then he goes ahead and stops performing those actions which do not lead to miseries and troubles. He even stops performing action which lead to pleasures of different sorts. Gradually he is able to reach final state.

Means to Attain Liberation

A man has to exhaust the old accumulated stock of merits and demerits by experiencing them. By means of right knowledge of the soul and reality, he also cultivates qualities like contentment, self-control prescribed by the scriptures. By means of these practices, by knowledge and mainly by arresting the further influx of karma, by denying any kind of experience of pleasure and pain one can ensure no further return of the soul into this world. When the soul is withdrawn properly from the world, it rests quietly in its own natural state. Liberation is thus according to Mimāmsā, a state of uttar painlessness as well as one of pleasurelessness. It so insists in the freedom from any kind of experience. 'In such a state the liberated soul rests in its own natural state, as he is free from pleasure as well as pain.'[60]

Thus the conception of liberation according to Prabhākara is not different from Nyāya-Vaiśeṛika conception of Mokṣa.

60. Ibid., 1.1.1, p. 130.

For both it is purely a negative state, a state, where there is no experience, neither pleasurable nor painful. The soul simply exists being free from all its accidental qualities including consciousness. Thus actions (obligatory, performed disinterestedly) and knowledge of the soul, both are recognised as the means to attain liberation by Pabhākara.

Kumāril Bhatta's Conception of Mokṣa

According to Kumāril Bhatta also liberation consists in the destruction of karma and the prevention of the production of future body. But unlike Prabhākara, Kumaril has explicitly stated his idea of Mokṣa in Śloka Vārtikā "for those who have understood the real nature of the soul, all their past karmas having become exhausted through experience, and there being no further kārmic residum left to wipe off, there comes no further body, as it is only for the experiencing of the reactions of past karmas that the soul is burdened with the body. Therefore the seeker of liberation should not do any such act as has been forbiddened or even what has been enjoined for certain purposes (as both these would bring about kārmic reaction which would have to be expiated by experience) but he should continue to perform the compulsory act as the omission of these would involve sin, which would have to be expiated by painful experience through a physical body."[61] According to Nyāy-ratnakāra, "liberation consists in the destruction of the present body and non-production of the future body."

Latter Bhāṭṭas differ in their opinion whether the soul enjoys bliss in the state of liberation or not. According to one view, there remains another kind of happiness which is the very nature of the soul. The happiness remains unmanifested in the empirical life but manifested in the state of liberation. The experiencing of his happiness requires mind. So according to these thinkers, there remains mind in the state of liberation. According to others, there is only a potentiality of happiness, which is the very nature of the soul. Thus the soul returns to

61. Pūrva Mimāmsā Indian History Sources : Dr. G.N. Jhā. pp.37-38.

its original state of liberation. The nature of liberation according to Prabhākara and Kumāril Bhatta is the same. It is purely a negative state because no positive entity is found eternal. They also agree that it is not merely the cessation of pain but also of pleasures because pleasures are always accompanied by pain. It is not a blisssful state as it is for Vedānta because according to Mimāmsakas bliss requires body for its enjoyment and embodiment is bondage. Hence it is not a blissful state.

Embodiment is the sole cause of bondage, according to Mimāmsakas. Freedom from embodiment is liberation. Embodied release is a contradiction.

Vedānta holds that knowledge can burn the old stock of Sanchita karma but Kumārila does not seem to agree with it in this respect. However, he holds that knowledge helps the prevention of the accumulation of the new karma and leads to the stoppage of further embodiment of the soul. He does not neglect the importance of knowledge for the attainment of release though he lays greater stress on duties.

The approach of Nyāya Vaiśeṣika and the Mimāmsā to the Philosophical problems is the same. That is why the conception of liberation of both is almost the same. The Pūrva-Mīmāmsā system does not believe in embodied release i.e. it does not believe in 'Jīvanmukti'. i.e. living liberation.

Chapter - III

CONCEPT OF MOKṢA IN VAIṢṆAVA-VEDĀNTINS, ŚAIVAS, ŚĀKTAS, PRASTHĀNATRAYĪ

Vaiṣṇava Āgamas

Bhedābhedavāda of Bhāskara

Bhāskara wrote a commentary on the Brahmasūtra called the Bhāskarabhāṣya about 900 A.D. He was the upholder of Bhedābhedavāda, or the doctrine that unity and multiplicity are equally real. Brahman is not an undifferentiated mass of pure consciousness, but possesses all perfections. The causal state of Brahman is called unity and its evolved condition is one of the multiplicity. Things are non-different in their causal and generic aspects and different as effects and individuals.[1]

According to Bhāskara, Brahman is divine personality. The world is not an appearance of Brahman. He admits 'Brahma-pariṇāmavāda' i.e. transformation of Brahman and tells that, 'the world is real and it is transformation of Brahman. It is essentially spiritual. Reality is both, one and many. Brahman is one in the causal state,in effect state it is many. Both cause and effect are real. Though Brahman is partless, he transforms himself into the world and remains unchanged in himself. He is immanent in the world and transcends it. He is neither absolute nor identity, nor absolute difference. Just as the sea is identical with its wave, which are its modifications and yet different from them, so Brahman is identical with the world, which is his modification and yet different from it.[2]

---

1. Indian Philosophy : Dr. S. Radhākriṣṇan, p. 670.
2. Foundation of Hinduism : Dr. Y.s. Shastri, p. 94.

He holds that the world of matter has real existence, though it is essentially of the same nature as Brahman. When matter acts on Brahman, it serves as a limting adjuncts in the form of body and senses and results in the rise of individual souls. Thus the 'jīvas' are Brahman itself but limited by the adjuncts of ignorance, desire and other actions. The individual self is naturally identical with Brahman while its difference is due to material limitations. The relation of jīvas and Brahman is illustrated by the analogy of sparks and fire. The life of Saṁsāra is based on the confusion between Brahman and Upādhi. Through virtue and piety, we can discriminte between the two, and then we are saved.

Liberation is identity of jīva with Brahman. The identity between them is essential. Their difference is due to the limiting adjuncts. Bondage of the jīva is due to desire for enjoyment of worldly objects and its liberation is on account of its realization of Brahman. Brahman, thus, as limtied by souls, suffers the miseries of the world and obtains liberation therefrom through a combination of true knowledge and performance of duties,i.e. jñānakarma-Samuccaya.

Bhāskara does not admit embodied release i.e. jīvanmukti, there is only disembodied release i.e. Videhamukti in which the finite self becomes one with the Absolute, and realises its infinitude and absoluteness.

TheViśiṣṭādavaita of Rāmānuja

Rāmānuja's school of philosophy is known as 'Viśiṣṭhādvaita' or 'Qualified Monism'. He was the chief exponent of Vaiṣṇava school of philosophy. Ramānuja was performed scholar and a keen Vaiṣṇava devotee. In the post-Śankarite period and among the successors of Śānkarite period and among the successors of Śaṅkara, Rāmānujācārya stands as the most prominent and powerful personality. He critised Śaṁkara's Māyāvāda with a great vehemence and established his own philosophy. He lived in the eleventh century. Yādava-prakāśa' was his guru at Conjeevaram. He was a great follower of Vaiṣṇavism, his creed was widely accepted and his influence on Vaiṣṇava community is very great.

He has written his commentary on Bhagavad-gītā, Vedārthasangraha and Gadyatraya. His commentary on Brahmansūtra is known as Śribhāṣya. Āgamaprāmāṇya, Siddhitraya and Gītārthasangraha by Yāmunācārya are equally important works of this school.[3]

According to Ramānuja there are three ultimate entities—God, individual souls and matter. God is the same as the Supreme Brahman of Upaniṣads. He is an organic whole. He contains plurality within itself. Individual souls and matter are the vital and inseparable parts of the 'Whole'. Brahman of Ramānuja is the abode of all auspicious qualities. Etymologically 'Brahman' means 'Brihattva' (greatness) and denotes, "The Highest Person' who is by nature devoid of all evil and is possessed of hosts of auspicious qualities which are innumerable and unsurpassed in excellence. The Brahman is both efficient and material cause of the univese. The universe being the creation of Brahman, can not be false in the sense in which Śaṅkara means. Individual souls are eternal like Brahman. They are not creation of Brahman like ether, fire, air etc. They come out of Brahman as sparks come out of a blazing fire. Thus the teachings of Vedānta are interpreted and developed by him in his own way. For him Brahman is not devoid of any quality but it possesses innumerable auspicious qualities and possesses its own personaity. Ramānuja usually describes the Brahman as Puruṣottama, Vāsudeva, Nārāyaṇa, Viṣṇu etc. For him Reality or Brahman is non-dual spirit, but not attributeless, distinctionless, homogeneous identity. He accepts internal differences within Brahman. Within the Brahman, there exists as part, the different unconscious material objects as well as many unconscious souls. As endowed with attributes, it is viśiṣṭa (that wich is qualified). Within the all-inclusive Brahman there are both unconscious matter and finite souls.

---

3. Foundation of Hinduism : Dr. Y.S. Shastri, p. 95.

The Brahman is the infinite substratum in which all changes take place Rāmānuja is Bhedābhedavādin and he admits the pariṇāmavāda-theory of causation according to which the effect is a transformation of the cause and both the cause and effect are real. The effect i.e. the world is not entirely different from the cause but it is the Brahman that differentiates itself into the two main modes as the soul and the matter. He holds that the individual souls and the inanimate objects exist as the body of the Brahman just as we have our bodies related to the soul. His Brahman is the highest God living in Vaikuṇṭha, qualified, acts as the source, sustainer, absorber of all manifestations which are finite.

The (Jīva) Soul is of atomic size and it dwells in the heart. It is not all pervasive. It is one and single for each person, extremely minute like an atom in size and dwells in the heart of each person—As a drop of sandle-ointment, although applied to one spot of the body only, produces a refreshing sensation extending over the whole body, thus, the self also, although dwelling in one part of the body only, is conscious of sensations taking place in any part of the body.[4] He does not identify the soul with knowledge like Śaṅkara, who had made the soul and knowledge identical. Rāmānuja, on the contrary, makes a distinction and maintains that the soul is different from knowledge and that the knowledge is the quality that belongs to the subject.

The souls are infinite in number and are of three classes : Firstly there are eternal or nitya jīvas which are never in bondage. Secondly the free or mukta souls which have already achieved their salvation and thirdly the bound souls or baddha jīvas which are caught up in the vortex of samsāra. The third type of baddha jīvas wander from life to life till they are released. Due to their past karmas they have to transmigrate from one life to another till it attains release.

Rāmānuja agrees with Śaṅkara and all others that bondage

4. Ātman and Mokṣa : Dr. G.N. Joshi, p. 521.

is due to nescience i.e. Avidyā. It is due to wrong identification of the soul with the body, senses and acit. Bondage is essentially an illusory phenomena, nescience can not exist anywhere but in the thinking mind and hence it has to be uprooted by right knowledge of the Brahman. Bondage is only imaginary and it has no permanent existence. Release or liberation is eternal because it is nothing but the face nature of the self and it is co-existent with Brahman. If the sense of separativeness is successfully overcome by meditating upon the real nature of Brahman, the soul losses all its worldly pain and sufferings. It realizes its real nature as pure and perfectly blissful. This is nothing but Mokṣa.

Attainment of Mokṣa or the realization of the essential identity of the soul with the Brahman is the aim of mumuksu. A seeker of liberation has to follow the three paths or yogas namely Path of knowledge-jñānayoga, Path of action-karmayoga and Path of devotion-bhaktiyoga. For the removal of bondage, knowledge also is necessary. Bondage arises out of nescience and neiscience can be destroyed only by the knowledge of Brahman. It can be removed by the study of Vedānta, karma i.e. performing duties, action, bhakti etc.

Ramānuja emphasizes on intense love for God. The grace of God leads to liberation. Htalks about utter self-surrender to the God.[5] The God is pleased by devotion and release the devotee from bondage. The liberated soul becomes similar to God, because like the God it has pure consciousness, free from imperfections. But it does not become identical with the God, as the finite can never be infinite. Through intense devotion the soul has a vision of the God, but it finally released only when his physical body perishes. He does not believe in embodied release i.e. Jīvanmukti, he believes only in videhamukti. So unlike Advaita Vedānta, there is no jīvanmukti in Viśiṣṭādvaita.

Thus Mokṣa of Rāmānuja is not like the Mokṣa of Śaṅkara, according to whom, the individual soul gets ultimately

5. Outlines of Hinduism : T.M.P. Mahādevan ppp. 152, 55, 56.

lost and dissolved into the Supreme Self, the brahman, but it is a positive state, in which though the individual soul losses its physical and subtle body, maintains, its separate consciousness and enjoys all the powers of Brahman except that of creation and regulation of the world. It also enjoys overwhelmingly the perfect intelligence, perfect bliss and all the innumerable auspicious qualities of the Supreme Self. It is a state of positive unending enjoyment of happiness in which the soul and the self are equal and remain inseparable.[6]

The Dvaita-Vedānta of Madhvācārya

Madhvācārya is also known as Pūrṇaprajñā and Ānandatīrtha. He was born in 1199 in a village near Udipi, of the South Canara district. He wrote commentary on Brahmansūtra, Bhagavadgīta and Upaniṣads. He is the exponent of the Dvaita or dualistic school of Vedānta. Jayatīrtha and Vyāsatīrtha are the most important of his successors. His main target of criticism was Advaita of Śaṅkara. Madhva was the champion of Dualism and accuses Śaṅkara of teaching the false doctrine of śūnyavāda of Buddhism under the cloak of Vedānta.

Madhva's Vedānta is called Dvaita-Vedānta because the concept of difference i.e. Bheda is the central point of this system. Madhva's Dvaita Vedānta is realistic pluralism. It is realism because it believes in the reality of the external world. It admits plurality of reals, so it is pluralism. It is a theism, since it accepts personal God who is the only independent reality, the other real beings are entirely dependnent on Him. Madhva advocates the reality of the five-fold differences between (i) God and Soul (ii) Soul and Soul (iii) Soul and Matter (iv) God and Matter (v) Matter and Matter. His bias for difference is so great that he advocats difference of degrees in the possession of knowledge and in the enjoyment of bliss in the state of liberation.[7]

---

6. Ibid., p. 555.
7. Foundations of Hinduism : Dr. Y.S. Shastri, p. 98.

jīva is totally dependent and is possessed of limited knowledge and limited joy. God is omnipotent and independent. His powers and perfection follow from his own nature. The Jīvas are infinite in number, atomic in size, the knowledge and bliss which are possessed by jīva are fragmentary as contrasted with those of God who possesses them to their perfection. The individual souls are the knowers, the doers of actions and also the experiencers of the fruit of actions. God enjoys sovereign power because he is independent. Madhva like Rāmānuja identifies God with Viṣṇu or Nārāyaṇa and calls him as Hari, Vāsudeva and tells that He can be known only by scriptures. He believes that God, soul and matter are the three entities, which are eternal and absolutely real, though soul and matter are dependent on God. God is perfect personality. He possesses infinitely good qualities. Existence, consciousness and bliss constitute His essence. He is the creator, preserver and destroyer of the univese. He is the doer of action directly or indirectly. All actions are caused by God, they are caused at the will of God. He is present in all the things, not in the gross form but in an extremely subtle form. God who is the impeller of actions of the jīvas, is also the awarder of the fruits of actions, because he is the Lord of karma. He has divine body and is transcendent. Laxmi is His consort.

The souls are of three kinds—(i) those who are eternally free i.e. nityamukta (ii) those who have attained freedom from samsāra and (iii) Baddha or bound. This classification is based upon the three guṇas or attributes. The sattva soul goes to heaven, the rajas revolve in samsāra, while the tāmas falls into hell.[8]

Madhva holds that everyting depends on God, God is responsible for every experience for every event of the world. He is responsible for knowledge as well as for ignorance, bondage as well as for Mokṣa. Everything happens at the will of God. Mokṣa occurs only with the grace of God. If God is

---

8. Indian Philosophy : Dr. S. Rādhākṛṣṇan Vol. II, p. 744.

pleased with His devotee, He shows His grace to him and gives him Mokṣa. Knowledge, Upāsanā and worship of Brahman makes individual soul possible to attain Mokṣa. Renunciation, enquanimity, self-control, acquaintance with love, self-surrender, attendance on Guru, Acquisition of knowledge from Guru, Reflection, Devotion, Love for God, Sympathy and love for others, performance of rites without desires, knowledge of Viṣṇu all these qualities are necessary for the attainment of Mokṣa. Thus Madhva stresses the need of cultivating all important moral virtues and yogic practices for the attainment of Mokṣa.

In Mokṣa the soul becomes bodiless, formless and completely free from the karmas and the subtle body. The soul enjoys the company of God (Īśvara) and lies at rest in its own nature, which is one of perfect knowledge and perfect bliss. It enjoys all the divine powers except that of creation and governance of the world. The soul retains its distinction even in Mokṣa. Difference i.e. Bheda is eternal. The soul thus remains distinct in Mokṣa and not lost in it. It enjoys positive Ānanda even in the absence of body.

The Dvaitādvaita of Nimbārka

Nimbārkācārya was a Telugu Brahmin, a powerful leader of Vaiṣṇanism. He lived sometime after Rāmānuja and prior to Madhva about eleventh century A.D. He wrote a commentary on Brahmasūtra called 'Vedāntapārijātasaurabha', as well as ten verses, 'Daśaśloki', elucidating his views on distinctness of jīva, Īśvara and jagat. His philosophy resembles to a great extent the philosophy of viśiṣṭhadvaita of Rāṁānuja. He advocates the doctrine of Dualistic non-dualism i.e. dvaitādvaita which stresses identity-in-differnece, i.e. Bhedābheda between God (Brahman) and the souls and the world. He points out that the world of souls and matter is both different and non-different from Brahman. He criticises both Advaita and Dvaita and reconciles the identity texts with the difference texts in his doctrine of Dvaitādvaita.

For him Brahman is the unitary and Ultimate Reality which stands as the source, sustenance and reabsorption of the whole world and souls. At the same time he rejects vivartavāda or illusoriness of the world, and the souls of Śaṅkara and agrees with Rāmānuja in holding that the Brahman as well as individual souls and the world are equally real. The Brahman is related to the world by the relation of Bhinnābhinna i.e. Difference and Non-diffeence. Brahman is the only independent reality. Brahman is saguna and not nirguṇa. Whenever scriptures mention Brahman as nirguṇa, it means that evil qualities do not find place in it. Brahman is eternal, omnipresent, omnipotent and omniscient. Brahman is personal God, the Gopālakṛṣṇa accompanied by Rādhā. He is possessed by excellent qualities and powers.

The jīva is atomic in size. It is the knower, the agent, the experiencer, limited in its knowledge and bliss, enveloped by karma caused by Avidyā, which is beginningiess but terminable, it is limited in its knowledge and powers and it suffers in the world. It is extremely minute, as minute as the hundredth part of the extemity of a hair, when divided into hundred parts. Though it is jñānaswarūpa, it is not intelligence itself, as Śaṅkara holds. Nimbārka maintains a difference between the two by regarding knowledge as an attribute of the soul. The soul is its possesser. Jīvas are many in number but they are of two kinds (i) those that are mukta and live in supremely blissful condition and (ii) those that are tied down to the circle of saṁsāra. The jīva suffers from bondage so long as it regards itself independent and identifies itself with Prakṛti. The soul attains liberation when it realises its dependence, on the Paramātman and also realises that it is in reality one with Him.

It attains liberation when it realises that the sense of its complete independence and difference from the Paramātman is false and such a sense is caused by Avidyā. It overcomes its bondage and Avidyā by the grace of God. The devotee of Paramātman who seeks liberation must know the nature of the

Being to be worshipped i.e. the nature of God being constituted of existence, intelligence and joy i.e. sat, Chit and Ānanda. The nature of the worshipper that it is an atom, possesing limited knowledge and joy and it is the servant of the Kṛṣna. The fruit of God's grace, the self-surrender and the giving up of all actions except the service of God.

Nimbārkācārya relies more upon the efficacy of self-surender to the Divine for the attainment of Mokṣa. He states that the seeker of knowledge of Brahman should necessarily cultivate the virtues of calmness and tranquility by means of indifference towards all objects of enjoyment, by practicing complete self-control over his desires and sense organs.

Though the soul is part of Brahman, it does not lose its individuality even in the state of release. It only realises its essential similarity to God. The liberated soul experiences the very Brahman which is possessed of glory and brilliance the soul enjoys the essence of it and becomes happy forever. The soul.becomes more powerful at the same time it is free from the affairs of the world. It enjoys the perfect and supreme bliss with Kṛṣṇa.

Nimbārka advocates the necessity of the performance of the Vedic duties of caste and stages of life so long as they help the attainment of wisdom, although afterwards they become unnecessary. He threw the doors of Mokṣa open to the persons of all casts and he simplified the way of attainment of Mokṣa by means of bhakti to God, that culminated into total surrender to God. He emphasized the importance of knolwedge along with karma, the religious duties to be performed with a feeling of complete faith, love and reverence for the Paramātman by means of love, fully selfsurrender to God.

Śuddhādvaita of Vallabhācārya

Vallabha was a Telugu Brahmin of south India. He migrated to north and developed the view of Viṣṇusvāmin. He belonged to the thirteenth century. He accepts the authority not only of the Upaniṣads, Bhagavadgītā and the Brahmasūtra, but also of the Bhāgavata Purāṇa. In his works, Aṇubhāṣya, Siddhānta-rahasya and Bhagavatatīkā suboddhini, he offers a theistic interpretation of the Vedānta, which differs from those of Śaṅkara and Rāmānuja. His view is called Śuddhādvaita, or Pure non-dualism. He declares that the whole world is real. The individual souls and the inanimate world are in essence one with Brahman. Vallabh admits that jīva, kāla or time and prakṛti or māyā are eternal existences. They are refered to the being of Brahman and have no separate existence. While Śaṅkara traces the world to Brahman through the forces of māyā, Vallabha holds that Brahman can create the world without any connection with such a principle as Māyā.[9]

Vallabha founded Kṛṣṇa-cutt, for him Kṛṣṇa is Supreme Brahman. Brahman is the highest independent reality and is identified with Śrikṛṣṇa. He is Puruṣottama, the Supreme person. He is sat, cit, ānanda i.e. existence, consciousness and bliss and sentiment (rasa). He is the abode of all good qualities. He is smaller than the smallest and greater than the greatest. The universe springs forth from Brahman as sparks spring forth from fire. Souls and matter are his real manifestations. They are his parts. Though eternal and unchanging, He be manifests himself through his māyāśakti, but this manifastation is not illusory, it is real. Vallabhācārya's view is neither Vivarta nor Pariṇāmavāda,it is something in between the two called Avikṛtapariṇāmavāda. The universe is not vivarta (appearance) because it is real manifestation. It is also not a pariṇāma for this manifestation does not involve any change or transformation. The universe is a natural emanation for Brahman, which does not involve any notion of change and is, therefore, called Avikṛtapariṇāmavāda.[10]

According to Vallabha, the jīva is atomic, extremely minute in size. It becomes pervasive so that it may experience the sensation all over the body even though it occupies one portion of it. It becomes pervasive by means of its intelligence

---

9. Indian Philosophy : Dr. S. Rādhākṛṣṇan, p. 756.
10. Foundation of Hinduism : Dr. Y.S. Shastri, p. 102.

which is its property just as the fragrance spreads over a great distance from the flower. The souls are many and eternal. In the soul the same Brahman is present but its bliss is concealed. The jīva is supposed to be governed by ignorance when it fails to understand its real nature characterised as saccidānanda and when it falsely identifies itself with the material objects like the body,senses,mind etc.Vallabha advoctes the realization of the truth of the text 'That thou art' for the attainment of liberation. Vallabha means in the real sense the identity of the two while it is only metaphorical in the case of others. The soul is other than and beyond the body, the senses,the manas, and the intellect.

The Jīvas are of two kinds (i) The saṁsārins (Worldly) and (ii) the muktas (liberated). The saṁsāri jīvas move through the circle of existences and are subjct to misery which is caused by the false knowledge that the body and the senses are the soul. This state of misery continues till it acquires the knowledge of Brahman,till it devotes itself to meditation and love of God. Such souls remain miserable so long as they are under bondage which is caused by Avidyā or ignorance. Pure souls are free from ignorance. The souls attain liberation by knowledge or by devotion.

Vallabha regards that liberation can be attained by acquiring sāyujyatā or penetration into the being of the Lord Kṛṣṇa by means of singular and intense devotion. Those worldy souls in whom there is no divine nature and evil qualities predominate are known as āsura jīvas. They always move in the circle of existences. They lack divine qualities. When God shows compassion to them and relieves them from ignorance, their defects are removed and they become pure souls. In them intelligence and bliss manifests fully.

The jīvas who possess divine nature are of two kinds (i) Maryādāmārgīya and (ii) Puṣṭimārgīya. The Maryādāmargīya jīvas are distinguished by their following the path of knowledge, karma, religious and moral duties, bhakti, concentrated

11. Vedānta : V.S. Ghate, p. 38.

devotion, yogic practices as prescribed by scriptures. The Puṣṭimārgīya jīvas solely depend upon God for his grace and favour and by that achieve the final state of Mokṣa.

The Puṣṭimārgīya develop ardent love for God whom they look upon as their lover. The devotion of this kind is known as Premābhakti. Mokṣa according to Puṣṭimārgīya consists in the attainment of the abode of the Divine enjoyment that are identical with the nature of the Puruṣottama, in the company of Puruṣottama Himself. It is the everlasting experience of the unfading eternal joy of the Divine which the liberated soul enjoys in company with the Lord Puruṣottama. The Maryādā bhaktas attain Sāyujyamukti which means the actual penetration of liberated soul in the being of Puruṣottama by means of knowledge, religious duties, bhakti and the yogic practices, moral virtues etc., to attain sāyujyatā which is their highest aim.

According to vallabhacārya liberation means ultimate release of the soul from the body. Pleased by the devotion of the soul, God takes him to his abode and there the soul enjoys eternal bliss of playing with God (Rāsalīlā). Highly pleased by devotion, God keeps the devotee near Himself to enjoy the sweetness of service.

The Acintya-Bhedābheda of Caitanya

Caitanya Mahāprabhu flourished in Bengal between 1485 and 1533. He was a very powerful and inspired Vaiṣṇava saints. He was also known as Lord Gauranga or Nimai. He was also called Kṛṣṇa Caitanya for he possessed the unique power of inspiring love and devotion for Kṛṣṇa in all those who happened to come in his contact. He himself lived for the whole of his life-time a God-intoxicated person and used to fal very frequenlty into transces as a result of his mystical experiences of communion with the Lord Kṛṣṇa.

Caitanya's system is called as Acintya Bhedābhedavāda or Unthinkable Identity-in-Diffeence, due to the unthinkable power of God. His philosophical teachings are found in a work called 'Daśamūlaśloka' ascribed to him and Caitanyacaritāmṛta of Kṛṣ-ṇadāsa, 'Satsandarbha' of Goswāmī. Govinda-bhāṣya on the Vedāntasūtra of Baladeva. Caitanya was a great devotee of Lord Kṛṣṇa, naturally Rādha-Kriṣṇa was the Supreme Reality for Him.

The God or Supreme Reality is determinate. Brahman is not a bare distinctionless identity He has infinite number of Śaktis which are supernormal and inseparable. He is the creator, preserver and destroyer of the world. Among finite powers, there are three main powers namely Svarūpāsakti, Māyāśakti, and jīvaśakti. They are also called Viṣṇuśakti, Avidyāśakti and Kṣetrajñāśakti. The Svarupaśakti is his internal power (antarangaśakti). The māyāśakti is God's external power (bahirangaśakti), which is responsible for the material universe. God's Jīvaśakti forms the essence of the finite souls, the world and the souls spring from the nature of the God. The world is a modification of Māyāśakti and souls spring from Jīvaśakti. The world is not an appearance of Brahman. It is real. The doctrine of Māyā is a false doctrine. The jīva is a conscious power of God, while the world is energy of the matter. Souls are related to God as sparks to fire or as parts to the whole.[12]

Caitanya practised and preached Madhurā bhakti which is essential for reaching God. He preached that faith and devotion are the power ful means to attain liberation. No other means except ardent and singular devotion to Kṛṣṇa can rid a man from the pains of the worldly life. His Madhurā bhakti is a synthesis of four kinds of emotions known as śānta rasa, dāsyaras, sakhya rasa and vātsalya rasa, and it becomes a unique fusion by the acquaintance with the true nature of God. The communion with God with the help of Madhura bhakti brings an unparalled peace with it. The soul enjoys everlasting fountain of love, joy, calm, freshness and enthusiasm of the Divine. When a devotee is fully satisfied by drinking God's love to the fullest extent he falls into swoon or in an unconscious state by totally forgetting his narow indviduality. Caitanya himself had such a long communion with Kṛṣna.

12 History of Indian Philosophy : Jadunāth Sinha, p. 721.

Mokṣa consists in the eternal experience of love. Souls in heaven realise their status as the servants of God, and are utterly devoted to him. Love is release. Bhakti is true mukti. Through it bondage to rebirth is broken and the soul attains to a status of equality with God, though it is never absorbed in God. Thus bhakti is the way to salvation. Study of the Vedas the Bhagavat-purāṇa and the like, is inculcated. Reverence for the Guru is a cardinal feature. No man or a woman is too low for the grace of God. Ethical virtues of mercy towards all creatures, humility, tranquility, freedom from worldly desires and purity of heart are necessary to attain God's grace.

Concept of Mokṣa According to Āgamas

(i) Vaiṣṇava Āgama
(ii) Śaiva Āgama
(iii) Śākta Āgama

The Āgamas are manuals of worship. Āgamas sometimes are called Tantras. They represent an attempt of the early religious thinkers of the Indian at making available to the common man, the wisdom of the Vedas. Significant development of Hindu thought is found in these Āgamas. In the post-Brāhman period performance of Vedic rituals and sacrifices became quite difficult due to their rigid orthodoxy. At this juncture Āgamas came for help to satisfy religious needs of the society. The Āgamas are usually known as Tantras. They prescribe earlier and less complicated methods of worship, suitable for all the classes and masses. Āgamas are held euqally important as Vedas. They are not opposed to the Vedas. Popular Hinduism is thoroughly influenced by the Āgamas. Even some of the Indian philosophical systems base their authority on the Āgamas. These texts prescribe the mode of worshipping deities, and prescribe a way of life in confirmity with the nature of that worship.[13]

There are mainly three kinds of Āgamas—Vaiṣṇava, Śaiva and Śākta. Vaiṣṇava Āgamas consider Viṣṇu as the Supreme

13. Foundation of Hinduism : Dr. Y.S. Shastri, p. 64.

God, Śaiva Āgamas believe Śiva as the Supreme deity and Śākta Āgamas treat Śakti or mother Godess as the Supreme Deity. It is an attempt to combine Theism with Absolutism that expressed itself in these three sects, Vaiṣṇavism, Śaivism and Śāktism. All these three sects have their own independent sacred literature.

Vaisṇava Āgamas

Vaiṣṇava Āgamas are called Pāñcaratra Āgamas as Saṁhitās. They are designated as Saṇhitās, showing their connection with the Vedic texts. Vaiṣṇava Āgamas mainly deal with idolatry, rules of temple architecture, worship of Tulsi plant (holy basil) and lay emphasis on path of devotion which suits to all. Viṣṇu and his consort Lakṣmi are the main deities of these Āgamas. More than one hundred and eight Āgamas with various commentaries, belonging to this sect shows its popularity. The Śātvatasaṁhitā.Pauṣkarasaṁhita Jayākhyāyasaṁhita Ahirbudhnyasamhitā, Vaikhānasasaṁhitā, Pratiṣtāsāra, Nāradapañcaratra are some of important works of Vaiṣṇava Āgamic Literature. These Āgamās are called Pañcarātra, because it is believed that these texts are revealed by the Suprme Lord Nārāyaṇa or Viṣṇu Himself in five days. Another belief is, that Āgama literature deals with fivefold duties of routine life in a day, such as abhigamana, upādāna, ijyā, svādhyāya and yoga, therefore it is called Pañcarātra. Yamunācārya in his work, Āgamaprāmāṇya tells that, 'the Pañcarātra-saṁhita' is as valid as Veda since both are derived from the same divine source—Nārāyaṇa or Vāsudeva.[14]

These Vaiṣṇava Āgamas lay down a way of life for their followers. They prescribe rules for the worship of the images and symbols in houses and temples. These rules deal mainly with knowledge meditation and concentration. The rules are connected with building of temples and installing idols, method of worship and daily rites. Abhigama means going to the temple of God and concentrating mind, speech and actions on Him. Upādāna means, collecting material for worship of the God.

14. Ibid., p. 65.

Ijyā is actual worship. Svādhyāya is study of scriptures. Yoga means meditation on the God. Vaiṣṇava Āgamas declare that—'From Brahmā to a blade of grass everything is the Kriṣṇa'.

Mokṣa

The Vaiṣṇava ācāryas like Bhāskara, Rāmānuja, Nimbārka, Madhva, Vallabha, Caitanya hold that liberation is identity of jīva with Brahman. The identity between them is essential. Their difference is due to limited adjuncts. Bondage is due to desire for emjoyment of worldly objects. Bondage of the soul means its confinement of this body. The cause of bondage is karma which springs from ignorance. The ignorance of the jīva is destroyed by the grace of God, which reveals the nature of Reality. Knowledge without devotion can not lead to liberation, but unflinching love towards the God, lead to liberation, without knowledge. The culmination of bhakti is surrendering self to the God.

It is Nārāyaṇa or Viṣṇu or Vāsudeva or Kṛṣṇa whose grace leads to Mokṣa. Utter self surrender to God is necessary. The liberated soul then becomes similar to God. As the soul is a part of Brahman, it does not lose its individuality even in release.

Bhedābhedavāda of Bhāskara, Viśiṣṭhādvaita of Rāmānuja, Dvaita Vedānta of Madhva, Dvaitādvaita of Nimbārka, the Śuddhādvaita of Vallabha, the Acintya Bhedābheda of Caitanya, all these Vaiṣṇava ācāryas do not agree with ācārya Śaṅkara's Advaita Vedānta.

Śaiva Āgamas

Śaivism is a powerful religious creed in the Southern India like Vaiṣṇavism. Both are the rivals of each other. The Vaiṣṇavas regard Viṣṇu or Kriṣṇa or Rāma, or Nārāyaṇa as the Ultimate Reality—Brahman. The Śaivas, like Vaiṣṇavas look upon Śiva as the Ultimate Reality and the cause and substratum of the universe. Śaivism prevailed in south India even before Christian era. It received a great access of strength from its opposition to Buddhism and Jainism, which it, along with

Vaiṣṇavism, overcame about the fifth or the sixth century after Christ. It elaborated a distinctive philosophy called the Śaiva Siddhānta about the eleventh centhury A.D.

Śaivism of Kaśmir was influenced by the Vedic conception of Rudra and the Rudra-Śiva cult of the Brahmaṇās, the Mahābhārata and the Śvetāśvatara Upaniṣad. Besides these twenty-eight Śaiva Āgamas especially the parts dealing with jñāna or knowledge, the hymns of the Śaiva Saints, and the works of the later theologians, form the chief sources of southern Śaivism.[15]

The Āgamas of Śaivism have traditionally been divided into four types namely, the Soma, the Lākula (or Nākula), the Pāśupata and the Śaiva. The Śaiva Āgamas are further divided into three groups, namely, the left hand Āgamas (vāma) of the Aghoras; Kāpilas and Kālamukhas; the right hand (Dakṣina) Āgamas of Kaśmir Śaivism or Trika system based on Svaccanda and other Āgamas and twenty-eight Āgamas of Śaiva Siddhānta. The Kāmika, Suprabheda and Vātula are highly authoritative texts of Śaiva Āgamas. The Vīrāśaivism of Karnāṭaka is based on Śaiva Āgamas.

Śaiva Āgamas are a treasure-house of information concerning philosophical doctrine, litergy, religious practice, architecture of the Śiva temples, sculpture of the images and art in general as well.

Śiva is looked upon a Supreme Lord. He is called Paśupati i.e. Lord of all beings. 'Paśu' means 'a living being' and 'Pati' means 'Lord.' Paśu literally means best and 'pasu' word is derived from paśa i.e. bonds. In other words all living beings are bound by fetters.[16]

Śaiva Siddhānta

The philosophy of the Śaiva creed is known as the Śaiva Siddhānta.

The Paśus whom the Lord Śiva rules are the invidual souls

---

15. Indian Philosophy : Dr. S. Rādhakṛṣṇan Vol. II, p. 733.
16. Foundation of Hinduism : Dr. Y.S. Shastri, p. 66.

having the atomic size and are self-conscious. The Paśus are all pervading, eternal, doers of actions having imperfect knowledge and powers compared to Śiva, but when they secure mukti, or emancipation after their fetters i.e. pāśas are removed, then they become omniscient, posses the real power of actions.

Pāśas are of four kinds (i) Mala or taint which hides the powers of knowing and acting of the soul, and is like the husk that envelopes the grain of rice (iii) karma—impression of deeds for the attainment of fruits (iii) Māyā—the material cause into which the world returns at the time of dissolution, and the world springs from it (iv) Rodhaśakti—the obstructive power of Śiva. It regulates the other three fetters and conceals the true nature of the soul. Śiva is possessed of this power and hence, He can create the world.[17]

The paśus attain mokṣa after they become free from the four fetters by means of sincere and singular devotion to God, Śiva. Śiva can be attained only by those who exercise self-control and dedicate themselves to Him exclusively and love Him wholeheartedly. The souls (paśus) become Śiva in emancipation, they become omniscient, and their powers also enhance, excepting that of creation. The paśus become eternally free from ignorance, karma and miseries of the worldly life. This is Mokṣa according to Śaiva siddhānta.

Kaśmir Śaivism

Kaśmir Śaivism is a well-known branch of Śaivism which itself is again redivided into two branches known as the Spandaśāstra and Pratyabhijñāśāstra. According to Spanda-śāstra, Māyā is Lord's śakti or power, which is peculiar and it accomplishes something that is impossible of accomplishment for any other agency. Māyā is the Tirodhānaśakti, the wonder working power. According to this school, God is independent and creates merely by the force of His will all that comes into existence. He makes the world appear in Himself. The individual soul also is identical with the Lord Śiva, but it does not perceive this identity with Śiva due to impurities which is of three kinds.

17. Ātman and Mokṣa : Dr. G.N. Joshi, p. 611.

The soul forgets its free and universal nature through ignorance and believes itself to be imperfect. When impurities (i) caused by the finitude of the soul (ii) caused by Māyā and (iii) resulting from actions are removed by a devotee and attains the vision of Śiva by means of the most intense contemplation of him. When this condition becomes stable, the jīva becomes liberated forever, and becomes the Supreme Self, the Śiva Himself. Thus spandaśāstra branch of Kaśmir Śaivism holds that Mokṣa consists in attaining identity with Śiva by the jīva and not simply attaining similarity with Him.

According to Pratyabhijñā branch of Kaśmir Śaivism, jīva is Śiva Himself concealed by Māyā. The identity of jīva and Śiva is forgotten due to ignorance. The Pratyabhijñā hold that the soul can become self or the jīva can become Śiva only when and only if it recognises its original identity with the self. The jīva can regain its lost divine nature by simply recognising (pratyabhijñā) that it itself is nothing else but the self. This is possible by (i) externally knowing the divine attribute of God Śiva and developing love for Him and (ii) internally by developing the consciousness that—I am the God Himself.It is necessary to recognise the Lord Śiva, as the essential nature of the jīva. It is absolutely necessary for the attainment of liberation. Thus recognition of one-self as God is the way of Mokṣa.[18]

Vīrāśaiva or Lingāyata Sect.

Vīraśaivism of Karnāṭaka founded by Basaveśvara in twelth century A.D. is also based on Śaiva Āgamas. it is also called Lingāyata on account of the fact that its followers wear a linga (symbol of Śiva) on their person. Besides Vedas, Āgamas and Purāṇas, the Vīraśaivas accept vacana of Kannaḍa mystics. The Vīraśaiva philosophy is known as Śaktiviśistād-vaita. According to this school, God is one and is qualified by His power (śakti). Śiva and śakti are inseparable. The individual soul is part (aṁśa) of para Śiva (the highest Lord or Śiva) in the sense that it precedes from Śiva, pertakes of his essence and finds final rest in Him.

18. Ibid., p. 614-15.

The souls ultimate goal is to realise union (aikya) with Śiva. In this union, the soul enjoys Supreme bliss. The final state of experience is called lingāṅga-sāmarasya, identity in essence between Linga (Śiva and aṅga part) i.e. soul. Paraśaiva is the Supreme Reality in Vīrasaivism, called sthala, which means abode or place. In Him the universe exists and in Him it merges.[19] The Vīraśaivas also admit the possibility of the living liberated soul (jīvanmukta) who are free from ego and manifest the Divine peace and joy while alive.

Śākta-Āgamas

Śākta Āgamas represent a mode of subtle worship and a fine philosophy induced by Upaniṣadic thought. In these Śākta Āgamas we find the quintessence of mysticism which is based on the Upaniṣadic doctrine of the unity of individual soul and the Supreme Reality. (Brahman-Śiva). These Āgamas are usually called Tantras. (manual of worship)

The entire Śākta literature is in the form of a dilogue between the Lord Śiva and Pārvati (His consort). It is believed that the revealer of some Śākta Tantras is Lord Śiva and of some other Tantras, revealer is Godess Pārvati. The Śākta Tantric Āgamas recommend worship of Śakti or the universal mother. Śakti-worship is the form of worshipping the Supreme Brahman as a female principle.

There are three sub-schools within the Śākta Tantra namely kaula, Miśra and Samaya. They have their own independent treatises. The Kaula group has 64 Āgamas with various commentaries. The Miśra group has eight Āgamas. These two schools emphasise external worship and their methods are mostly used to acquire material power and prosperity. Samayā-cāra is suppossed to be efficient method of worship of the great Goddess. This school represents well-known Śakti, Worship, viz. Śrividyā. In these Tantric texts, Śakti or universal mother is a synonym of Brahman of the

19. Foundations of Hinduism : Dr. Y.S. Shastri, p. 67.

Upaniṣads. She is known as Mahāśakti, Lalitā, Tripurasundari and by other names. She is indeterminate as well as determinate. At the transcendental level, she is indeterminate, while at the phenomenal level she is determinate. At the highest level, she is the highest knowledge, beyond space and time, and by its nature, existence, bliss and consciousness. She is the only 'Reality' which pervades the whole universe. This world is the expansion of this universal consciousness, and therefore real. According to this Tantra the Upaniṣadic impersonal Brahman is conceived as goddess in the saguṇa form for religious worship. The attitude of forgiveness, compassion and love are more natural to mother than to the father. Thus Brahman is personified form, becomes affectionate mother, worshipped as Mahātripurasundari or Shri Lalitā[20]

The jīvas, the individual souls are the evolutes of Śakti, the Māyā. They are finite centres of consciousness and are produced out of ignorance. They are not unreal, but their reality or experience is limited. They cannot have the full experience of Brahman. The experience of the jīva, though limited, is real, but out of ignorance it thinks that it is the agent or doer of actions. It does not recognise Śakti as the real agent of all actions. Its bondage is therefore caused by ignorance.

Knowledge of Śakti is the road to Mokṣa, which is dissolution in the blissful effulgence of the Supreme. Mokṣa according to the Śāktas, is not dependent on the observance of the external forms of religion like recitation of hymns, sacrifices, and other rituals. What is necessary for the attainment of Mokṣa is the knowledge for jīva, that it is an evolute of śakti, and is, in essencial identical with it, and bhakti or devotion along with the physical and mental discipline help it takeit on the word. The Śāktas admit the possibility of jīvanmukti and believe in the doctrine of the transmitration of the soul.

20. Ibid., p. 68.
21. Indian Philosophy : Dr. S. Rādhākṛṣṇab Vol.II, p. 736-737.

CONCEPT OF MOKṢA IN PRASTHĀNATRAYI UPANIṢADS

The Upaniṣads form the concluding portions of the Veda, and are therefore called Veda-aṅta, or the end of the Vedas, a denomination which suggests that they contain the essence of the Vedic teachings. They are the foundations on which most of the later philosophies and religions of India rest.

The word 'Upaniṣads' come from upa-ni-sad that means sitting down near the teacher to receive instructions. It gradually come to mean what we receive from the teacher, a sort of secret doctrine or 'rahasyam' (secret). Sometimes it is made to mean what enables us to destroy error, and approach truth. Śaṅkara in his introduction to the Taittiriya Upaniṣad. says : knowledge of Brahman is called Upaniṣad because in the case of those who devote themselves to it, the bonds of conception, birth, decay etc. become unloosed, or because it destroys them altogether, or because it leads the pupil very near to Brahman, or because therein the highest God is sea-ted.[22]

Upanisads express the restlessness and striving of the human mind to grasp the true nature of reality.

Upaniṣadic view of Mokṣa is that, it is the highest state of rapture and ecstasy, a condition of Ānaṅda, where the creature as creature is abolished, but becomes one with the creator, or more accurately realises his oneness with Him. We can not describe this perfection adequately. We use symbols. The nature of eternal life is a condition of Ānaṅda or freedom, a state of joyous expansion of the soul, where heaven and earth are felt to flow together.[23]

There are many passages where the individual is said to become one with the highest. To the ancient thinkers the aim of philosophycal inquiry was, mainly the attainment of Mokṣ-

---

22. Indian Philosophy (Centenary Ed.) Vol I : S. Rādhākrishnan pp. 137-138.
23. Ibid pp. 286-287.

a, mukti, amṛtatva. Knowledge was the mean to the end of self-realization. Highest knowledge itself is understood as liberation.

Muṇḍaka Upaniṣad (ii.2.2)—Praṇava is the bow, the Ātman is the arrow and the Brahman is said to be mark, It should be hit by one who is self-collected, and that which hits becomes like the arrow, one with the mark, that is Brahman.

Muṇḍaka Up. (iii.2.5)—Having attained him, the seers content with their knowledge, their purpose accomplished, free from all desire, and with full composure, having attained the all pervading Ātman on all sides, ever concentrating their minds, enter into everything.

The liberated soul feels his oneness with God so intensely that he calls himself the creator of the world.

Taittireīya Upaniṣad (iii)—" I am the food, I am the food —eater, I am the subject, I am the object, I am the two together. I am the first-born, the destroyer of the world also. I am the sunlike light, I am the centre of the world, of immortal Gods."

Muṇḍaka Upaniṣad (3. 2. 8)—"As the flowing rivers disappear in the sea loosing their name and form thus a wise man, freed from name and form goes to the Divine Person, who is greater than the great. Here it is described the nature of the union of the individual soul with Supreme Self".

Kathopaniṣad (4.15)— As pure water poured forth into pure, becomes the very same, so becomes the soul (Ātman). There is no one to experience and no one to get the experience. There the experience becomes the experiencer and the experiencer the experience. They are one and the same thing. The knower becomes the known and viceversa.

In fact Mokṣa is not a thing to be attained, because individual soul is eternally free but it is misunderstood as body and mind due to nescience. As soon as Avidyā disappears, one experiences non-duality. It is a state of perfection. In liberation the enjoyer and the enjoyment are one and the same. It is a state of blissfulness, fearlessness. Eternal peace and joy rule there. The quantity of bliss is immeasurable and indescribable.

Taittireya Up. (2. 9)—He who knows the bliss of that Brahman from where all speech with the mind turns away, unable to reach it, he fears nothing.

Thus it is a state of eternal bliss and contentment. It is a state of perfection and fullness. The bliss that is experienced in this state is of highest magnitude. it is beyond imagination and comprehension. Only those who actually experience it, know it. This is a state of liberated soul. It is self-realization, spiritual attainment. It is like reborn in a divine life. It is a life of different category, in which the individual no more belongs to the human kingdom. It is a complete transformation of the human soul into the divine.

It is repeatedly said in the Upaniṣads that by the vision of the one Ātman, by overcoming the distinction, one attains perfections and immortality, and at last it becomes to this....."One who knows Brahman, becomes Brahman" (Mun. Up. 3. 2. 1.) ब्रह्मवेद ब्रह्मैव भवति।

That means one enters into identity with the Brahman which is also the Ātman. To be the Brahman means to possess all the charactaristics of Brahman, which are eternity, immortality, blissfulness, consciousness, incorporeality, infinity, tranquility, purity, freedom and luminosity. The individual acquires all these attributes and enjoys the ultimate state of liberation. So according to Upaniṣads Mokṣa means knowing one's self as Brahman. Experiencing that Ātman and Brahman are not separate. At this point dvaita disappears and advaita is experienced.

"ब्रह्मविदाप्नाति परम्।" (Taitt. Up. 2. 1. 1)

"The knower of Brahman attains the highest".

"तरति शोकमात्मवित्।" (Chan. Up. 7. 1. 3)

" The knower of the self goes beyond grief".

"अभयं वै जनक प्राप्तोऽसि।" (Br. Up. 4. 1. 4)

"Oh! Janaka! You have attained fearlessness".

"एतावद रे खलु अमृतत्वमेति।" (Br. Up. 4. 5. 15)

" Oh! (Maitreyi), thus for goes immortality".

"तमेव विदित्वा अतिमृत्युमेति।" (Sve. Up. 3. 8, 6. 15)

"Knowing It (Self) one goes beyond death".

"यत्र वाऽस्य सर्वमात्मैवाऽभूत्।" (Br. Up. 2. 4. 13)

"When Self only is all this".

Thus who realises Brahman attains Mokṣa which is the highest aim of life.

BHAGAVAD GITA

Bhagavadgītā literally means 'Lord's Song'. It is one of the great religio-philosophical classics of mankind. It is in true sense a crest–jewel of the Hinduism. It is part of the Bhiṣmaparva of Mahābhārata. It contains seven hundred verses which are divided into eighteen chapters. Almost all great ācāryas (Vedāntic teachers) and even Hindu exponents like Shri Aurobindo to Mahātmā Gāndhi Have written commentaries on it. The Gītā is considered as the essence of Upaniṣadic teachings. It is said that all Upaniṣads are like cows, Lord is the cowherd and Arjuna is calf, wise is the enjoyer, and the nector like milk is the Gitā itself. Being the essence of Upaniṣads, Gītā is elevated to the status of Upaniṣads.

Upaniṣads were taught in a peaceful setting of but the setting of Bhagavadgitā was the war field. This is the gospel given to Arjuna by Lord Kriṣṇa on the first day of Mahābhārata war. Arjuna's chariot, whose charioteer was Lord Kriṣṇa himself, was on the central ground between both the parties. Arjuna decided not to fight. In this moment of weakness, Lord Kriṣṇa conversed with him to cheer him up and philosophical truths behind all actions. This gospel of him is Gītā and the essence of Upaniṣads.[24]

It is a book conveying lessons of philosophy, religion and ethics. It is not looked upon as a śruti, or a revealed scrip-

24. Foundation of Hindusim : Dr. Y. S. Shastri pp. 45-47.

ture, but it is regarded as smṛiti, or a tradition. Yet if the hold which a work has on the mind of man is any clue to its importance, then the Gitā is the most influential work in Indian thought. Its message of deliverance is simple. While the rich could buy off the Gods by their sacrifices and only the cultured could persue the way of knowledge, the Gitā teaches a method which is within the reach of all, that of Bhakti, or devotion to God. The poet makes the teacher the very God descended to humanity. He is supposed to address Arjuna, the representative of man, at a great crisis in his life. The teacher is the favourite God of India, who is at once human and divine.[25]

Arjuna is found in the first chapter of the Gitā is what the mystics call the dark night of the soul, an essential step in the upward path. The further stages of Illumination and realization are found in the course of the dialougue. From the second chapter onwards we have a philosophical analysis. The essential thing in man is not the body or the senses, but the changeless spirit. The mind of Arjuna is switched on to a new path. The life of the soul is symbolised by the battlefield of Kurukṣetra and the kauravas are the enemies who impede the progress of the soul. Arjuna attempts to recapture the kingdom of man by resisting the temptations and controlling the passions. The path of progress is through suffering and self abnegation.

Krisṇa stands for the voice of God, delivering his message in thrilling notes warning Arjuna against dejection of spirit. The opening chapter shows great insight into the heart of man, conflicts of motives, the force of selfishness and the subtle whisperings of the evil one. As the dialogue proceeds the dramatic elements disappear. The echoes of the battle field die away, and we have only an interview between God and man. The chariot of war becomes the lonely cell of meditation and the corner of the battle field where the voices of the world are stilled, a fit place for thoughts on the supreme.[26]

The message of the Gītā is universal in its scope. Lord

25. Indian Philosophy (Vol I) : Dr. S. Radhakrisnan pp. 419-20.
26. Ibid–Page 521.

Kriṣṇa addresses no sect, establishes no school, but opens the way to all the winds that blow. He sympathises with all forms of worship.

Gītā teaches different yogas such as jñānayoga, karmayoga, bhaktiyoga, Rājayoga etc. Gitā makes room for the persons of different temperaments. The person in whom willing predominates, for them there is karmayoga, in whom feeling predominates, there is bhaktiyoga, in whom thinking predominates for them there is jñānayoga. The word yoga is derived from the root 'Yuj' which means to unite or to join. Yoga therefore denotes union of the individual soul with the Brahman, the Universal soul or God. The Goal of human life is thus, the realization of the self with entire existence and these yogas are the different paths to attain realization of the self. i. e. Mokṣa. Each individual is free to choose his own path suitable to his temperament. These paths are not opposed to each other but they are complementary to each other. Lord Kriṣṇa preaches each path in detail in each adhyāya.

Jñānayoga–Jñānayoga is the path of knowledge, path of wisdom. It is a path of intellectual discrimination between the real and the unreal, the noumen and the phenomenal. Knowledge means realization of one's own self. This path is very difficult, suitable only for intellectually and spiritually advanced souls, the Gitā places jñānayoga on higher level by stating that 'there is nothing purer than knowledge' (IV.38). Just as the kindled fire reduces to ashes all fuel, O' Arjuna so does the fire of knowledge reduce to ashes all action. (IV.37) He whose undertakings are all dissociated from desires and intentions, and whose actions are consumed in the fire of knowledge, him the wise proclaim the man of true insight (IV.19)

ब्रह्मार्पणं ब्रह्म हविर्ब्रह्माग्नौ ब्रह्मणा हुतम्।
ब्रह्मैव तेन गन्तव्यं ब्रह्मकर्म समाधिना।।२४।।

"For, (his) act of offering is Brahman, the oblation is Brahman, offered by Brahman itself, into the fire which (too) is Brahman (hence) the goal to be attained by him through concentration upon actions (i. e. act–Relation) which are themselves Brahman, is Brahman it self. (IV.24)"

Therefore, by the sword of knowledge cleave asunder this doubt that is born of ignorance, and betake this yoga and stand forth, O Descendent of Bharata (IV.42)

Karmayoga : –Karmayoga is a path of action. Karma here means niṣkāmakarma i. e. performance of action without any attachment to its fruits. Gitā emphasizes that............O, Pārtha, perform your own duties. "You have a rightful title to action, but only to action, never at all to its fruitions. 'Let not the fruits of actions be your motive, nor let your attachment be to inaction." (2. 47)

"Steadfast in yoga, do your actions, O Dhananjaya relinqushing attachment and becoming even-tempered (alike) in attainment or in non-attainment, (such) evenness is called yoga." (2. 48)

समत्वं योग उच्यते।

All the actions whatever their purpose may–should be the offering to the divine. 'Yoga' in Gitā means 'the art of doing action'. 'योग कर्मसु कौशलम्।' Evenness of mind is the sole essence of 'Yoga', wherein are naturally unity of vision (illumination) and action. Yoga is balance of mind.

Bhaktiyoga :— Bhaktiyoga is a path of love for God. It is easy to follow for one and all. There is a deep and emotional relationship between the individual and the God in Bhaktiyoga. To put in other words 'Brahman' can be attained through 'Bhakti' i.e. through emotional relationship with God. Twelth adhyāya in Gitā narrates the importance of 'Bhakti', as to bhaktiyogin attains 'Brahman'. What type of devotee is dear to him? Lord Krisna says, "Ever–content, the yogin possessing self-control and a firm resolve, having dedicated all mind and buddhi i.e. intellect devoted to me, is dear to me" "From whom the world shrinks not in disgust and who from the world shrinks not in disgust one that is free from elation, intolerance, fear and disgust, he that is (thus) devoted to me is, dear to me."

"He who is the same to enemy or to a friend, same also in honour and dishonour, who is the same in cold and heat, weal and woe, and is free from attachment." "To whom all alike is censure as well as praise, and who maintains silence, being content with anything whatsoever, not cumbered by a home firm in judgement, he is devoted to me, is dear to me. (XII 14, 15, 18, 19).

Rājayoga :—Rājayoga is another path to reach the universal self prescribed by Gitā. Mind is fickle and unsteady by nature. Stability of mind is necessary in reaching the goal of realization. Rājayoga is means of acquiring control and stability of mind. Continuous practice and complete detachment lead to the stability of mind.

Gitā proclaims that everyone has the spark of divinity within him. It says that 'the Lord resides within the hearts of each individual'.

ईश्वरः सर्व भूतानां हृद्देशेर्जुन तिष्ठति। (XVIII-61)

Therefore let your mind be full with my presence, fill your soul with My love, bow down to My divine being in all. He alone worships me by true sacrifice, who burns up all desires in his unswerving devotion to me. Here indeed I tell you the innermost secret, with your heart full of my divine being you shall reach my eternal home, and realise that treasure of eternal bliss (IX 34). Indix your mind in me alone, make your intellect find its home in me, and then ever afterwards you shall doubtless within me alone (XII 8)

Gitā teaches that the soul is indestructible, eternal, unborn and uncaused. Neither the weapons are capable to destroy it nor fire can burn it.

नैनं छिन्दन्ति शस्त्राणि। (II-44)

It is only physical body which is destroyed and not the soul. This soul is liberated from the circuit of birth and death when the eternalness of soul is known. Any path prescribed in Gitā leads to liberation.

In the eighteenth the last chapter of Gitā, Lord Krişṇa narrates Mokṣa Saṅyāsa Yoga. By the word saṅyāsa, he means not the renunciation of action but the desires and expectations implied in the actions. This is true saṅyāsa. In this chapter again action without attachment is repeatedly evaluated.

Mokṣa can be attained by them who have pure mind, control over their senses, who serve others, do there actions without expecting fruits of them, who preach the message of Gitā to others, who are jñāni, who surrender themselves to Almighty etc.

Gītā also makes room for gradual release (Krama-mukti). The yogins who have known Brahman are successively led (after departing from body) by fire God, the Gods presiding over daytime, the bright fortnight and six months of the northward course of the sun and finally he reaches Brahman. (VIII 24)

It is declared in Gitā that even womanfolk, vaiśyas, and those born of sin (pariah) can attain supreme goal provided thay take the refuge in the Lord. (IX-32).

The fact is that, hardly one amongst thousands of men strive to achieve the goal, and among the (few) seekers that strive (for it) hardly one in reality gains the knowledge of Me as I am. (VII-3).

Thus Gītā is at once both, the art of living and the science of Reality. This is declared in the colophone which occurs at the end of every chapter. "Gītā is at once the root, the flower and the fruit of ancient Vedic and Upanisadic culture".[27]

सर्व धर्मान् परित्यज्य मामेकं शरणं व्रज।
अहं त्वा सर्व पापेभ्यो मोक्षयिष्यामि मा शुचः।। (XVIII. 66)

Relinquish all (So called) duties, take my shelter, I shall 'release' you from all the sins whatsoever.

27. The Bhagavadgitā and its background : S. C. Roy, p. XIII.

BRAHMASŪTRA

Upaniṣads are the end of the Vedas or Vedānta, in respect of study the Upaniṣads come last. Some of the Upaniṣads are also known as Āraṇyakas, or the forest treaties. The Upaniṣads were many in number (112)[28]. They were developed in different vedic schools at different time and places. The problems discussed and solutions offered presented differences in spite of a unity of general out look. The need was felt, therefore, in course of time for systematizing the different–teachings so as to bring out the harmony underlying them. Bādarāyaṇa's Brahmasūtra undertakes this task. Brahmasūtra is also known as Vedānta–Sutra. It discusses in four chapters (a) the coherence (समन्वय) of the upaniṣadic teachings (b) their non contradiction (अविरोध) in relation to established theories and logical rules. (c) the means of realization (साधना) and (d) the fruit achieved.[29]

In the samanvayādhyāya, it is shown that Brhman is the Ultimate Reality, and that all the upaniṣadic texts have Brahman for their purport. Objections raised by the critics are met in the second adhyāya and the untenebility of non-vedāntic theories, is exhibited. The third Sadhanādyāya deals with the means for attaining Brahman and the forth phalādhyāya discusses the fruit of Brahmavidyā. It refutes the dualistic sāṁkhya opinion which conceives puruṣa and prakriti as two independent realities and hold that it is one Brahman which is the Transcendental Reality.

The Brahmasūtra though, adopts systematic procedure to explain philosophical teachings of the Upaniṣads, it is written in the form of aphorism, which are very difficult to understand without the help of commentaries. Each Ācārya (master) has interpreted Brahmasūtra in his own way and thus we have the Vedāntic schools of Śaṅkara, Bhāskara, Rāmānuja, Madhvācārya, Nimbārkācārya, Vallabhācary, Chaitanya, Swāmināraāyaṇa and many others.[30]

According to Bādarāyaṇa Brahman is the efficient and

28. Hist. of. Indian Phi. VOL I : S. N. Dasgupta., p. 28.
29. An Intro. To Indian Phi. : S. Chattergy and D. Datta., p. 348.
30. Foundation of Hindulims : Dr. Y.S. Shastri–p. 88.

the material cause of the universe. The universe is originated from it, is sustained by it and merges in it after destruction. So Brahman is the cause of origination, sustenance and destruction of the universe. It does not require any external help in creating the universe. It creates the universe out of itself as a spider weaves a net from within itself so it is both material and efficient cause of the universe. The Brahmasūtra commences with the aphorism.—

"अथा तो ब्रह्मजिज्ञासा।"

"Then, therefore, the desire to know Brahman.' The individual soul is atomic in size. It is of the nature of intelligence. It is not only knower but also knowledge at the same time. It is a real agent-though its activity is dependent on Brahman. The individual soul is a part of Brahman. The soul is neither absolutely identical with Brahman nor absolutely different from it.

The meditation on Brahman leads to knowledge. In the liberation of him, who has acquired such knowledge of Brahman by meditation, there is delay only so long as his body does not fall. But the action done by him after the attainment of knowledge does not bind him, because his notion of agency is destroyed. He has realised that he is not an agent but it is only God who is acting through him. But he has to exhaust the prārabdha karmas by experiencing the fruit of them. After the fall of the body, starts the spiritual pilgrimage of the soul to Brahmalok. The first thing which happens is that his function of speech merges into mind. Then gradually the function of all the organs merge into mind. Then mind itself merges into prāṇa, prāṇa then merges into individual soul. After this the soul of the knower of the Lord enters the nerve suṣumnā and then goes out of the body and takes the path of Gods.[31]

The complete enumeration of the stages of the path of Gods is mentioned by Swāmi Vishveshvarānanda as follows—First the deity of the flame of fire, then the deity of the

31. Brahmasūtra Śānkarabhāṣya IV 2. 1 to 7.

day, the deity of the bright half of the month, the deities of the six months, when the sun travels to the north, the deity of the year, the deity of the world of Gods, the deity of the air, the sun, the moon, the deity of the lightening, the world of varuṇa, the world of Iṅdra, then prajāpati and finally to Brahmaloka.[32]

The soul in the state of liberation merges into Brahman, but even there it does not lose its individuality (for Śaṅkara this is only vyāvahārika satya, pāramārthika satya is that the question of merging does not at all arise. It simply realises itself as non different from Brahman) it enjoys fellowship with Brahman. It becomes satya–saṅkalpa i.e. it can get anything by simply desiring the world of fathers, by its mere desire they come to it. it is the master of itself. It can have or not the body at its own discretion.[33] If it wishes it can create many bodies for the sake of enjoyment at the same time. It can share in all the powers of the Lord except the poweres relating to creation etc. of the Universe and impating grace. Thus the soul does not become purely identical with Brahman, but becomes only similar to it. This similarity is confined to the enjoyments and not with regard to the powers. Once the soul has reached the world of Brahma, it should not fear that it will have to come again to this world.[34]

This summery accounts is accepted by Rāmānuja and Nimbārkācārya without reservation. But Śaṅkara, Madhvācārya, and Vallabha can not accept it without certain important reservations. According to Śaṅkara the definition of Brahman as the 'Creator' etc. shows only the taṭasthalakṣ aṇa of Brahman, and its svarūpalakṣaṇa is found in scriptural passages as truth, knowledge and infinity.[35] The creation etc indeed to describe only the lower Brahman or Isvara. No activity can be ascribed to Higher Brahman. According to him Ātman and Brahman are identical. The activity of Ātman is also

32. Ibid., p. 197.
33. Ibid., V 4. 12.
34. Ibid., IV 4.22.
35. Taittariya Up : II 1.1.

according to Śaṅkara not the natural attribute of Ātman but due to limiting adjuncts of the body, sense organs etc, the spiritual pilgrimage shown above is only applicable to the worshiper of Saguṇa Brahman (Īsvara) and not the nirguṇa or Higher Brahman. According to him the prāṇas of the knower of nirguna Brahman do not depart from the body at death. They merge in him only. Again a realized (vidvat) becomes one with Brahman when liberated. In fact he was ever one with Brahman but only comes to know this fact in liberation. Thus in the realization of liberation there is a change only phenomenally or epistemologically and not ontologically.

An impartial study of the sūtras will make it clear that there is no room for the Māyāvāda of Śaṅkara in it. In nirguṇa Brahman of Śaṅkara, the falsity of the world and the absolute identity of jīva and Brahman appear to be unknown to Bādarāyaṇa. Really speaking Brahmasūtra of Bādarāyaṇa represents that stage of Vedānta when it was free from all the dogmas of later Vedantin schools—but Bādarāyaṇa seems to be more theistic than an absolutist like his commentator Śaṅkara.[36]

---

36. Vedānta : V.S. Ghate, p. 169.

Chapter - IV

CONCEPT OF MOKṢA AND JĪVANMUKTI IN ADVAITA VEDĀNTA (UTTAR MIMĀṀSĀ)

(I) PRE-ŚAṄKARA ADVAITINS-GAUḌAPĀDĀCĀRYA

Gauḍapāda is the first systematic exponent of the Advaita Vedānta. He is reputed to be the teacher of Śaṅkara's teacher, Govindayati, and is said to have lived about the beginning of the eighth century or at the end of the seventh. It is said that Gauḍapāda wrote a commentary on the 'Uttaragītā'. The central principles of the Advaita Philosophy, such as the orders of Reality, the identity of Brahman and Ātman, Māyā, the inapplicability of causation to ultimate reality, jnāna or wisdom as direct means to Mokṣa or freedom, and the inconceivbility of absolute nothing are set forth in his Māṅḍukyopaniṣad Kārikā.1

This work is divided into four chapters. The first called the Āgama, explains the text of the Mandukyopaniṣad. Gauḍapāda tries to show that his view of reality is sanctioned by Śruti and supported by reasons. The second chapter, called Vaitathya explains by means of arguments the phenomenal nature of the world, characterised as it is by duality and opposition. The third part establishes the Advaita theory. In the last part, called Alātaśānti or Quenching the Firebrand, there is a future development of the Advaita position regarding the sole reality of the Ātman and the relative character of our ordinary experience. As a stick burning at the end when waved round, quickly produces an illusion of a circle of fire (alātacakra), so it is with the multiplicity of the world.[2]

1. Indian Philosophy : S. Radhakrishnan, Vol. II, p. 452.
2. Ibid., p. 453.

Gauḍapāda is the philosophical progetior of Shri Śaṅkara. There is a famous stanza in which we get the hirerarchy of Ācāryas in the line of Shri Śaṅkara and his disciples. We find that hierarchy starts with Lord Viṣṇu from whom in the following sequence the great knowledge has reached us.—Shri Nārāyaṇa, Vasiṣṭha, Śakti, her son Parāśara, Vyāsa, Śuka Gauḍapāda, Govindapāda, Shri Śaṅkara, Padmapāda, Hastāmalak and Trotakacārya.[3]

Gauḍapāda's age was surcharged with Buddhism. Buddhism was politically respected, favoured by scholars and dogmatically accepted by the common man. Buddhism was in air at that time. So Gauḍapāda lived at a time when Buddhism was widely prevalant. Naturally he was familiar with the Buddhistic doctrines. He accepted some of them which were not in conflict with his own Advaita. His liberal views enabled him to accept doctrines, accosiated with Buddhism and adjust them to the Advaitic design.[4]

In his Kārikā on Māṇḍukya Upaniṣad, he elabortely discusses his thoughts on the illusoriness of the phenomenal world, non-duality of the individual self with the Supreme Self, Ānandamaya state of liberation etc. In Muktikopaniṣad we have the briefest and yet the most glorious review on Māṇḍukya wherein it is said

> 'Manḍukya alone is sufficient for an aspirant to attain liberation'[5]

Gauḍapāda elaborately discusses his thoughts on Mokṣa in his Kārikā. He believes that the word 'AUM' is all this. All that is past, present and future verily, is 'AUM'. That which is beyond the three periods of time is also, indeed 'AUM'[6]

---

3. Maṇḍukya Upaniṣad : Gauḍapādakārikā : Swami Chinmayānanda, p. 288.
4. Advaita of Śaṅkara : Dr. S.G. Mungal, p. 133.
5. (Manḍdukyam ekam kevalam Mumukṣunām Vimuktaye) M. Up. : Swami Chinnayanand, p. 14.
6. Ibid., p. 22.

Just as millians of mud pots are all in a sense nothing but mud, so too there is a Divine Eternal Factor out of which the pluralistic world has arisen. The pots have been born from mud, are sustained in mud, again, must after their distruction, go back to be the mud, so too the pluralistic world of phenomena has no other substance for their stenance than the truth. In Truth they exist and to Truth they return when they are withdrawn from their state of manifestation. That fundamental truth, the Ultimate Eternal, the all-pervading conscious principle, is nothing but 'AUM'. It is the spiritual centre behind everyting. 'AUM' is Brahman. So all this is verily Brahman. This Ātman is Brahman. This Ātman has four padas i.e. quarters.[7]

'AUM' not only indicates the spiritual centre in the individual, but also represents at once the total Spiritual Reality behind the delusory concept of the world of plurality. The names and forms constituting the world are mere superimpositions upon all pervading consciousness. Vaiśvānara is its prathama pāda, whose sphere of activity is waking state, who is conscious of the external world of objects. Taijas is the second pāda which is related to the experiences in dream, sphere of activity is dream-state. The third quarter (pāda) is prajnā whose sphere is deep sleep. The fourth pāda is 'Turīya'. It is discribed as Adṛṣtam, avyavaharyam, agrāhyam, alakṣaṇam-------i.e. it is not that which is conscious of the internal subjective world, not that which is conscious of the external world, nor that which is conscious of both, nor that which is a mass of consciousness, not that which is simple consciousness, nor it is unconscious. It is unseen by any sense-organ, not related to anything, incomprehensible by mind, uninferable, unthinkable, undescribable, essential of the self alone, negation of all phenomena, the peaceful, all bliss and non-dual. This is what is known as the fourth pāda (Turiya).

---

7. Ibid., p. 25, 27 (Sarvam hi etad ayamātmā brahma so ayamātmā catuṣpād) ll2ll.
8. Ibid., p. 66 (Agam Prakaraṇa verses 1, 2, 7).

सोऽयमात्माऽध्यक्षरमोङ्कारोऽधिमात्रं पादा मात्रा।
मात्राश्चपादा अकार उकार मकार इति।।८।।

The same Ātman is again AUM from the point of view of the syllables. The AUM with parts is viewed from the standpoint of its sound or letters. The quarters are the letters and the letters are querters. The letters here are A, U, M.

The waking-state-ego i.e. Vaiśvānara is compared with the letter 'A' in AUM. It is superimposed upon 'A' sound of AUM, for the purpose of meditation.

The dream-state is compared with U sound in AUM. Taijas is considered to be superior to the 'waker'. It comes in between the two states, so is letter 'U' in AUM.

Prajñā, whose sphere of activity is deep sleep in 'M', the third letter of 'AUM'.[9]

The sound letter 'A' helps the meditator to attain a well developed waking state personality (i.e. viśva). The meditation on U attains a well-developed Tejasa (mind and intellect) and he who meditates on M attains prajñā. In the soundless, there is no attainment. This soundless part of 'AUM' is the goal of a meditator.

The partless 'AUM' is the soundless aspect—the silence that must necessarily be between two successive 'AUM's.

That which has no parts, the soundless, the incomprehensible, beyond all the senses, the cessation of all phenomena, all blissful and non-dual AUM, is the same as the Ātman. He who knows this, merges his self in the Supreme Self, the individual in the total.

युञ्जीत प्रणवेचेतः प्रणवो ब्रह्म निर्भयम्।
प्रणवे नित्ययुक्तस्य न भयं विद्यते क्वचित्।।२५।।

--Soak the mind with the roar of AUM. Identify the mind

9. Ibid., pp. 93 to 100.

with the sound of AUM, AUM is Brahman, the ever fearless. He who is always unified with AUM shall know no fear whatsoever.—

प्रणवं हीश्वरं विद्यात् सर्वस्य हृदि संस्थितम्।
सर्वव्यापिनमोङ्कारं मत्वा धीरो न शोचति॥२८॥

—know AUM to be Iśvara, the Lord, ever present in the minds of all, the man of discrimination realising AUM as All-Pervading does not ever grieve.—

अमात्रोऽनन्तमात्रश्च द्वैतस्योपशमः शिवः।
ओंकारो विदितो येन स मुनिर्नेतरो जनः॥२९॥

—One who has known AUM, which is moraeless and of multiple morae (i.e. AUM which is soundless and of infinit sounds) and which is ever peaceful because of negation of all duality in it, is the true sage, none other—

He who thus understands Omkāra, not the mere AUM sound, but the Truth signified by AUM, is the real man of refection, the real sage of wisdom and God-realization.

However intelligent he may be, however well respected in his time and place, none can claim Godhood merely because of his erudite scholarship,fiery expressions, eloquent writings or attractive pedagogue on Vedantā theories.

He alone is a sage who has rediscovered himself to be nothing other than the truth, the All-pervading and Eternal knowledge Absolute, indicated by the sacred syllable AUM.[10]

Thus Gauḍapāda emphasizes the truth that 'AUM' is Brahman itself and the Ātman is Brahman. He believes that the world we experience is illusory in nature and the knower of the world of delusions is Ātman. Ātman is self-luminious, through the powers of its own delusion (Māyā) imagines in itself by itself all the objects, and its, individual experiences both in the worlds, outside and within. It alone is the knower of the objects

10. Ibid., 111, 117, 119.

so created (page 139). It is the conscious principle of life in us. It is the very Ātman who is the seer. It is the knower of all the external sense objects of the world and the internal flow of different thoughts, emotions ideas etc. It is in the light of consciousness that we get our world of dreams illuminated. It is the light of Ātman that illumines for us the gross external light of the day in our waking state. It is all pervading Ātman conditioned by the mind and intellect in the ego-centric jīva, i.e. Individual Self. The individual self-superimposes its own likes and dislikes, loves and hartreds and thousand other emotions upon its own projection. If this 'outward gaze' is removed which is the cause of delusion and if the 'inward gaze' is attained, it comes to the state of self-awarness where the consciousness becomes as it were, consciousness itself. This is a state of Self-Realization and this is the process by which the pluralistic world of delusion can be folded out of cognition. He who realises Ātman, realises that it is the Universal Self which resides in this world of multiplicity and yet it is different from it.

स्वप्नमाये यथा दृष्टे गन्धर्वनगरं यथा।
तथा विश्वमिदं दृष्टं वेदान्तेषु विचक्षणैः॥३१॥

—As the dream of the illusion, or "the palace city of the fairy Morgana" are seen is the sky, so is the universe viewed by the experienced Vedāntins.—(31)

In fact the world of phenomena is nothing but delusion superimposed upon 'Reality'.

न निरोधो न चोत्पात्तिर्न बध्दो न च साधकः।
न मुमुक्षुर्न वै मुक्त इत्येषा परमार्थता॥३२॥

—There is neither dissolution, nor birth, neither anyone in bondage, nor any aspirant for wisdom; neither can there be any seeker for liberation, nor any liberated as such. This alone is the Supreme Truth.—

तत्वमाध्यामिकं दृष्ट्वा तत्त्वं दृष्ट्वा तु बात्ह्यतः।
तत्त्वीभूतस्तदरामः तत्वादप्रच्युतो भवेत्॥३८॥

—Having known the truth internally within the body as well as the same Truth externally in the outer world, he becomes one with the Reality, and thereafter derives his pleasures from it and never does he deviate from Truth.—This is a true realization. True realization is at once the realization of the divinity that is lying concealed in the individual himself.

Gauḍapāda calls this self-realization as the 'Aspaśayoga'. This term is taken from the Buddhistic literature.

अस्पर्शयोगो वै नाम दुर्दशः सर्वयोगिभिः।
योगिनो बिभ्यति ह्यस्मादभये भयदार्शिनः।।३९।।

—This yoga, called the Aspaśayoga (the touch-of-the-untouch) is hard to be attained by all seekers. Yogins are afraid of this path, for they feel frightened in That, where alone one can experience true state of fearlessness.

Some critics complain that the term 'Asparśa-yoga' is borrowed from Buddhistic sccripture by Gauḍapāda, and that there is no equivalent term in the scriptural texts, but this criticism is not acceptable. In Bhagavadgītā we get the evidence of it in the fifth adhyāya (21, 22, 27) wherein Lord has used the word 'aparśa' meaning mind's contact with the eternal objects through the sense organs. Gauḍapāda must have taken this expression from his knowledge of Gītā. So a samsārin who lead a finite life of sorrows is a 'sparśa yogin'. The God-man who is living the finite life of Eternal Truth is 'Aspaśa Yogin'.

Gauḍapāda proves this fact by giving the example of firebrands. He compares the pluralistic world with the different patterns created by the firebrand. As the firebrand with a glowing tip being swirled in different directions and create an appearance of different patterns. Ignorant children make a serious enquiry as to how and from where these patterns are created but the wise and educated know that those patterns are not real one and are caused by merely a glowing tip of a firebrand. These various shapes and patterns are compared with the world of phenomena. One whose mind is not still for him the world of phenomena appears to be true. One whose mind is still, knows that these patterns are illusory. It is only a glowing tip of firebrand which create these patterns. When firebrand is steady, all the forms get absorbed into the glowing tip. So one who realises the self for him self alone is real, the pluralistic world of phenomena is no more, it is absorbed in the Supreme Self.

One who has realised Brahman, there is nothing for him to accomplish or to achieve. The state of realization of Brahman is a state of perfect knowledge of identity between his individual self and the Total Self. He surrenders himself to the Supreme Self. It is merging of an individual at the feet of the 'whole'. Such a sage lives in eternal bliss. This is the goal of human life. It is a state of complete detachment from our body, mind, intellect and there by rediscovering ourselves to be our eternal nature, the Sacchidānanda Self.[11]

असंगोऽहम् असंगोऽहम् असंगोऽहम् पुनः पुनः।

सच्चिदानन्द रूपोऽहम् अहमेवाहमव्ययः॥

Unattached, Unattached, Unattached, am I Again and Again, of nature Asbsolute-Existence-Knowledge-Bliss am I. I am that. I am—the Eternal, the Immortal, the imperishable.

This state is nothing but jīvanmukti. A Jīvanmukta is a Living God on earth, a sthitaprajñā. Gauḍapāda accepts jīvanmukti. In his Kārikā, a jīvanmukta is denoted by different names such as Muni, Tattvavid (knower of Reality), Mukta, Buddha (Enlightened) Sanātana (ever the same) and the state of liberation as dukkhakśaya (destriction od misery), Uttamasukha, Abhaya, Veetaśokam, Niścalāsthiti, Sāmya, Akṣ ayaśānti, Sarvajñātā (omniscience), Āvarāṇacyuti etc.

Thus Manḍukya-kārikā of Gauḍapāda (or Āgamaśāstra) is considered to be the first available systematic treatise on Advaita-Vedantā. So his name in the history of Advaita Vedānta

11. Ibid., pp. 265, 283, 349, 422, 192.

remains as a great pioneer who convinced the followers of Upaniṣads that their path was sound. Ācārya Śaṅkara was rightly drawn to his teachings and found in him the spring of eternal life.[12]

(II) ŚAṄKARĀCĀRYA

Introduction :

In all the Vedāntic schools, Advaita Vedānta of Śaṅkarācharya has the highest respect and importance in the minds of Hindus. He was a great philosopher among the Indian thinkers in intellectual eminence. He was relentless in his criticism of other systems. He was zelous and untiring in the exposition of his own sytem of 'Absolute Idealism.' The intellectual atmosphere was surcharged with his 'Absolute Idealism' in his time.[1]

Before the rise of Śaṅkara's Advaita, the onslaughts of Buddhism on Hinduism was great and Hinduism based upon the Śrutis was facing a critical position. Śaṅkara constructed a new system of philosophy based upon the Veda and the Upaniṣads, which could meet the challenge of Buddhism and reconciled them with the Upaniṣadic thought. He counteracted the influence of Buddhism by admitting their important features and at the same time by refuting them on rational grounds.[2] There are different opinions about the period of this great philosopher but generally it is held that Śaṅkara was born in 788A.D. and died in 820 A.D. He was a Namboodari Brahmin. His ancestors were famous for their Vedic learning.

Even as a child Śaṅkara was genius. At the age of seven Śaṅkara was master of all Vedas, Vedangas, Itihas, Puranas etc. His guru was Govindapād, the pupil of Gauḍapāda. Govinda taught him the main principles of Advaita system. In the

---

12. Gauḍapāda : A study in early Vedānta, T.M.P. Mahādevan.
1. History Of Indian Phi. : Jadunath Sinha, p. 461.
2. Ātman and Mokṣa : G.N. Joshi, p. 443.

childhood, he was impressed with the mistery and importance of life and had an early vision of beauty and holiness. Before he learnt the ways of the world. he rejected them and became a sanyasin. A pure flame of truth burnt within him.

According to Śaṅkara, it was Advaita Philosophy alone, which could do justice to the truth of the conflicting creeds and so he wrote all his works with the one purpose of helping the individual to a realization of the identity of soul with Brahman, which is the means of liberation from saṁsāra. In his wondering from his birthplace in Malabar to the Himalayas in the North, he came across many places of worship and accepted all those, which had in them the power to elevate man and refine his life.

Though Śaṅkara was a pupil of Govindapāda and guided by Gauḍapāda, he was an independent thinker. He has not devised something which is absolutely neval and original. He has written commentaries on the Brahmasutra and the main Upaniṣads, which are independent from his own point of view, and reconciled them in a logically harmoneous system which is of great importance. He has supported his arguments with the help of Śrutis. He has not expounded his view separately but his view is automatically expounded through his various commentaries. A consistent system emerges from his commentaries. He had the originality of interpreting the sutras in a logically harmoneous ways.

The whole Philosophy of Śaṅkara is summarised in only a half verse *i.e.*

ब्रह्म सत्यं जगान्मेथ्या जीवो ब्रह्मैव नापरः।

Brahman alone in real. The world of plurality and multiplicity is false and ultimately the Ātman and the Brahman are non-different. The Philosophy of Śaṅkara is described as Advaitism which means that Ultimate Reality is single, non-dual, It is that which has no second. The Ultimate Reality is only one and that is the Absolute-Brahman.

The word 'Brahman' is derived from the root 'Brh' which means 'to grow'. So Brahman is that which 'grows'. It grows and extends so much that it embraces the whole of the existence in all the times. It is supreme, perfect, and Absolute Reality.

Brahman is the only substratum and root cause of all things of the past, present and future. All things are born out of it, they exist or subsist in it and by it and ultimately return to it at the end. It is the supreme and unchanging principle of existence which acts as the final abode and support of all things and phenomena.

According to Śaṅkara, Brahman is the highest transcendental truth. It is the ultimate Summum Bonum of human efforts and the basis of knowledge. It is existent, beginingless and unchanging. It is the highest Knowledge. One who knows, Brahman knows everything in the world. Knowledge of Brahman is eternal truth. Highest Knowledge itself is Brahman. The distinction between the highest knowledge, Knower and Known disappears in the highest state. Brahman is Sat-chit and Ananda.

This very Brahman is known as Ātman or self according to Śaṅkara. He accepts fully the Upanisads saying :

'All this is Brahman'

सर्वं खलु इदं ब्रह्म।

'I am the Brahman'

अहं ब्रह्मास्मि।

'soul is Brahman'

अयमात्मा ब्रह्म।[3]

Śaṅkara identifies the soul with Brahman, 'The Ātman is Brahman' There is one eternal, universal consiousness. This is the only, ontological Reality. The eternal, pure intelligent, free, omnipresent, perfect, immutable, blissful and self-luminious

3. Br. Up. pp. 14-10

Brahman is our real self. The Muṇḍaka Upanīṣad compares individual souls with the rivers and Supreme Self to the ocean and says that both being constituted of water are identical in nature but different in form and as the river becomes one by loosing itself in the ocean, so the individuals loose their entire individual separate nature by being one with the self Upaniṣads also compare the individual soul to the sparks coming out of blazing fire and Supreme Self to the blazing fire and points out that they are formed of the same nature as that of the general fire.[4]

Thus according to Śaṅkara the individual self is the Supreme Self itself, neither less nor more. Both are the same.

Bondage :

It is due to Avidyā that the individual self thinks it to be different from the Supreme Self. Avidyā is the cause of bondage. Due to bondage the individual self thinks itself-to be a body, mind, senses etc. Bondage is due to ignorance. Avidyā generates the psychological organism, which limits the universal consciousness and makes it appear to be the individual self. Avidyā is intellectual knowledge infected with the duality of subject. Avidyā is removed by intuition (Vidyā). Avidyā is bondage. Vidyā is liberation.

Liberation is higher knowledge

The Ātman is eternally liberated, but its intrinsic freedom is not known, because it is covered by Avidyā. Liberation is not an achievement but a discovery. It is like a discovery of a prince brought by a shepherd of his sovereignty when his identity is disclosed to him.[5] When Avidyā is destroyed its intrinsic, pure, divine nature is known, even as the real nature of rope is known when the illusion of snake is destroyed.

Integral experience uproots Avidyā, which is the root cause of saṁsāra. It destroys rebirth due to the conjunction of the Ātman with the mind, body, complex, which is caused by Avidya.[6]

---

4. Mundaka Up. II. 1. 1
5. S. Bhasya Br. Up. Ii. 1. 20
6. S.B. G. XIII. 28.

Karma theory : (Law of Karma)

Śaṅkara repeatedly speaks of Avidyā. Avidya is the cause of desire. Desire takes the form of attachment and aversion These are called faults or affliction. They are due to intellectual disorder. They are the springs of action. They produce actions (karma). Actions are either rightous or unrightous. Rightous actions produce merits. Unrightous actions produce demerits. These actions or karmas produce their fruits in the shape of joys and sorrows, in the present life or in the future life. They are the cause of rebirth or transmigration. Until the merits and demerits i.e. karmabeeja are completely burnt up by the integral experience of the Brahman, the wheel of birth and death continues to revolve. When they are exterminated by the true insight, transmigration ceases and Mokṣa is realised.[7]

Mokṣa :

Liberation of the Ātman consists in its own true nature, realizing the fact that it is absolutely one and non-different from Brahman. That is all what liberation is for Śaṅkara. As we do not attain, modify or produce a rope in overcoming of an illusion of rope-snake, so also the realization of liberation which is the overcoming of the cosmic illusion of Brahman-world is not an attainment or modification or production. It is rather discovery of the fact existing since eternity. Therefore it is described as too near and also far away. Being our very nature it is too near to us, but at the same time being covered by the mask of Avidyā, it is far away from us.[8] Mokṣa is the realization of Brahmānubhava, is not an event in space and time. It is also not realised with causality.

Mokṣa is Kūtastha-nitya i.e. immutably eternal. There are certain things which are changing at every moment, yet maintain their identity. The river, for instance is called pariṇāma-nitya. There are other things like mountains which do not

7. S. B. Taittiriya Up. Intro.
8. S. B. Īśa Up. 5.

undergo any change are called kūtastha-nitya. According to Śaṅkara. Brahman is devoid of any change and modification It is Brahmānubhava, Brahmānaṅda, Brahmanubhūti. It is devoid of any sort of change. It is kutastha-nitya.

There are some Śruti passages which declare that Brahman is self-luminions. Even the sun, the moon etc. get their power of luminosity from Brahman. Mokṣa is the realization of such Brahmanhood. So it is also self-luminious. The illusion of a rope appearing to be snake can not be overcome by any means other than knowledge. Similarly, Mokṣa is not possible by any means other than knowledge of Brahman thinks Śaṅkara.

Knowledge the only means to Mokṣa :

Śaṅkara emphatically maintains that it is knowledge and knowledge alone which is the means of Mokṣa. Highest knowledge is Mokṣa itself. The opposition of knowledge and action is like the opposition of light and darkness. The fruits of meritorious or sinful actions are happiness and misery respectively, but the only fruit of knowledge of Ātman is Mokṣa itself. Action and devotion may be useful indirectly in preparing the ground for the dawn of knowledge by way of purifying the mind but they are not directly the means of Mokṣa e.g. the illusion of rope-snake can not be overcome by action or devotion. Similarly it is not possible to overcome the cosmic illusion which is the samething as to realise 'Mokṣa' by action or devotion.

The function of knowledge is not to produce or create something new. It just removes the darkness of ignorance and in the light of which Ātman realises that it is non-different from Brahman. Thus Śaṅkara takes great pain to prove that liberation is not related to action.

A Liberated person :

One who has known himself as non-different from Brahman has noting to do, in the sense that there is no compulsion on him from the outside. He is now beyond good and evil. The liberated who was so far servant of moral law, becomes master of it. Moral activities are not destroyed but they transform into spontaneous activities. He is free from the struggle of 'is' and 'ought' which is at the root of moral activity. Moral activities become the very nature of a liberated. Dr. Radhakrishnan writes "These great men go on doing their daily work diffusing virtue as the star diffuses light and the flower perfumes without being aware of it.[9]"

He is above morality or becomes a supra-moral. So morality is the prerequisite for the attainment of liberation "Moral disciplines though necessary as aid to jñāna.......but are not enough for Mokṣa."[10] Morality is an ornament for him (S. B. 1-1-4). The world of plurality and multiplicity is not annihilated in the state of liberation of a liberated person. The wrong notion of world is annihilated. It is realised that the world of plurality is like a dream.[11] "The world of plurality does not disappear but is lit up by another light" It is finding the necklace on the neck itself. Similarly the Ātman was ever one with the Brahman and therefore liberated since eternity. As soon as one realises this fact, wrong indentification comes to an end and one becomes liberated.[12]

Śaṅkara also makes room for the religious minded people. He provides them Īsvara who is the highest appearance of nirguṇa Brahman and to whom devotion can be paid. His Īśvara serves the same practical purpose including normal and religious as that of other Vedāntins. The fate of a worshiper of the lower Brahman is the same in all respects as that of the liberated of Vaiṣṇava-Vedāntins. But all this is real from Vyāvahārika point of view. It is real so long as we are under the power of Māyā. When ignorance is removed, it is realised that everything experienced so far, including morality, religion, Īśvara and onés own similarity with Him was like a dreamt last night.

9. History of In Phi. : Dr. Radhakrishna, p. 229-230.
10. A study of Śaṅkara : N.M. Shastri., p. 168.
11. M. Karika II. 31.
12. In. Phi. : Dr. Radhakrishnan Vol. II., p. 584.

Jīvanmukti :

Śaṅkara is a staunch advocate of 'Jīvanmukti' i.e. liberation while embodied. Even after attaining the knowledge of the Brahman and transcending all distinctions of the world by knowing it to be illusory, an individual may continue to have his bodily existence in the world. Such a liberated person is known as 'Jīvanmukta.' Śaṅkara admits the possibility of a Jīvanmukta who has realised in the fullest degree the nature of the self by entering into communion with it. He has transcended the realm of relativity and so detached himself completely from the effects of the adjuncts of the world upon him by cultivating disinterestedness in all things, but still his physical body has not disappeared. Jīvanmukta continues to live in the body though he has discarded nescience (Avidyā) by the philosophical knowledge. The body does not immediately drop away even after the nescience is annihilated by the knowledge of the Brahman. e.g.trembling due to fear continues for sometime even after the false knowledge of the snake is terminated. Śaṅkara gives two analogies to prove the liberation while living. Firstly he says The potter's wheel keep on revolving though the stick that gives it movement is detached from the wheel, similarly Avidyā's impressions cannot be immediately and abruptly brought to an end though it itself is annihilated. Secondly he argues on the analogy of double moon vision.

As a man seeing two moons instead of one due to his eye disease, continues to see even after he is convinced that there is only one moon and not two. Similarly even after realizing that the unreality of the world appearance, the saint continues to witness the illusion of the world appearance, though internally he remains unaffected by it.[13]

'Jīvanmukta' lives in this world but is not of the world. He takes part in the worldly activities but is not affected in any way by them. To a Jīvanmukta the world exists only as a mirage in a desert. The external world is only an appearance to him

13. B. S. Śaṅkarabhasya IV. 1, 15, 19.

and so he shares no joys and no griefs of the world. He is sthitaprajña. He is neither happy nor unhappy. He is neither good nor bad. For him there are no desires, no passions, no fear, no anger, no joy, no grief, no envy, no love, no hatred. He is steady and permanently fixed in his own nature. He is a living God on earth ; an immortal among mortals, a perfect among the imperfects, an infinite among finites.[14]

Śaṅkara being a staunch advocate of Jīvanmukti asks, "Had there been no Jīvanmuktas, there would have been no teachers of knowledge". He boldly asks "How can one's own intimate experience of Brahman-Knowledge, while still embodied can be contested by anybody else?"[15]

Śaṅkara is the only Advaitin who emphatically believes and maintains that Jīvanmukti is possible. The conception of Jīvanmukti has remained controversial point among other Vedantins and also among some followers of Śaṅkara. It is criticised on the basis that body is said to be the effect of Avidyā or karma, while liberation is the knowledge of Brahman, when Brahmajñāna has dawned, ignorance should be removed with all its results, i.e. the body should fall down immediately. How is it possible that light (of Brahmajñāna) and darkness (the body i.e. effect of Avidyā) should exist simultaneously? Jīvanmukti is, on this point is a contradictory conception according to the opponents. They ask is it possible that sun may rise and the darkness with all its results, may also not disappears?

But in fact the above and similar to this objection against the conception of Jīvanmukti can not stand, because they are based on misunderstanding. Really speaking it is not action 'Perse', which is Avidyā, and cause of bondage. It is only Sakāma Karma action which is caused by unfulfilled desire, which is cause of bondage. Lokmanya Tilak writes....."gross or lifeless karma by itself does not either bind or release anybody. A man is bound by karma as a result of his hope for

---

14. Atman and Mokṣa : Dr. G.N. Joshi, p. 499
15. S. Bhasya IV. I. 15.

fruits or by its own attachments, and that when this attachment has been got rid off, a man stands released, not withstanding that he may be performing action by his external organs".[16] The other Vedāntins overlooked this truth, and in order to be free from the expected inconsistencies arising from the continuation of the body with Brahmajñāna they rejected the conception of Jīvanmukti itself.

Dr. Nalinikant Brahma thinks that the conception of Jīvanmukti is very valuable for Advaita Vedānta.

It should not be rejected at any rate According to him "the attempt to save consistency of the Vedāntic position by denying Jīvanmukti is like curing the disease of a patient by destroying his vitality. Jīvanmukti is crowning achievement of Advaita and can never be rejected on any basis."

Reason is not competent to establish or overthrow the possibility of the state Jīvanmukti. It is beyond its sphere. It is the sphere of 'anubhuti' or 'spiritual experience'. These experiences are inherited from our ancient sages. They are preserved in the form of Upanisads. Śaṅkara gives very important place to Śrutis in his pholosophy. Often it is said that Śruti is 'Pratyakṣa'[17] By saying this he means that it records the intimate spiritual experiences of the sages. Being the experienced, experiences of the sages, Śruti is valid in itself like the pratyakṣa or anubhava.

"Philosophy may be said from the standpoint of Advaita Vedānta to have its beginning and end in 'revelation'. It is impersonal to begin with and personal in the end. It begins with Śrutis and end in anubhūti, thereby making the spiritual heritage of mankind a personal possession."

So to reject Jīvanmukti is to reject Śrutis and Smṛitis and to reject the most vital part of human experience.

Śaṅkara establishes that liberation is possible by knowledge alone, neither by action nor by knowledge combined

16. Gītā–rahasya Lokamanya Tilak, p. 438.
17. Dr. Nalinikant Brahma : Hindu Sādhanā, p. 189.

with action. For Śaṅkara knowledge is not the dry intellectual knowledge. It is the illumination in the light of which everything is seen to be Brahman.

"Jñāna for Śaṅkara is not a cold study of the books but a warm hearted striving to realise the truth, which when turned towards a personal duty becomes bhakti".

He had the heart of a devotee.

Personality of Śaṅkara was a archetype of a combination of action, devotion and Knowledge. What is necessary according to him is one should have a self surrendering attitude of a devotee towards the fruits of action, while doing an action, a sādhaka should surrenders the fruits of all his actions to God. It is only then that one realises the ideal of niṣkāmakarma.

Śaṅkara's own life was a perfect example of jñāna bhakti and karma-yoga. His great contribution to Vedānta Philosophy, the great task performed by uniting India culturally, his writings, an important part of which consists of stotras to the different deities, is an ideal life of a Jīvanmukta.

VIDEHAMUKTI

The final disappearance :

The finally liberated soul is called Videhamukta, the bodiless liberated soul. In a Videhamukti the gross as well as the subtle body comes to an end. All the three karmas i.e. Sanchita Karma, Sanchīyamana karma and prārabdha karma come to an end. They are totally exhausted. The soul becomes completely free from rebirth. It ceases to revolve the wheel of Karma. This is the final state of liberation. The individual soul is lost for ever in Brahman. It is one with Brahman and enjoys to the fullest extent the perfect bliss of the Brahman. Nothing exists for it outside itself. It experiences Brahman.

The state of liberation is thus one that transcends all spacial temporal and mental limitations. It is indefinable, indescribable. Our language falls short to its description and hence nothing can be compared with it, hence uncomparable. It is unique in itself. It can not be known by reason or logic but knowing the soul which is freedom in itself is by intuition (sākṣatkāra), by entering into direct union or identity with it. Such a direct experience is immediate when the individual soul and the Brahman loose their distinctions. In the direct and immediate intuition of the knower, the known and Knowledge become united into one single whole. Mokṣa is thus a state which has to be actually experienced. It is supersensuous. All external means of Knowledge prove insufficient and inadequate to know it. It is properly known only by becoming it.

Śaṅkara maintains that people of all casts can attain liberation irrespective of sex differences. However he recognises the traditionally sanctioned Varṇāśrama-dharma and holds that the high cast people can attain it from the study of the scriptures and the others by bhakti. He nevertheless allows Mokṣa to any person of any cast, any sex and any race if the Knowledge of Brahman downs upon him even without any external means. We will see Kramamukti and Sadyomukti briefly. Advaita Vedantā also believes in these two besides Jīvanmukti and Videhamukti.

Kramamukti

Advaita Vedānta recognises four types of muktis. Kramamukti is one of them krama means gradual or step by step and mukti means deliverance. Kramamukti falls to the lot of those who worship Saguṇa Brahma and are possessers of lower knowledge (i.e. Saguṇa vidyā also known as Sāndilya Vidyā).[18] Due to their incapability or failure to attain the saving knowledge in this life, such souls pass initially to the world of Saguṇa Brahman (also known as KāryaBrahman) or Hiraṇyagarbha by the way of Gods or the Devayāna. The things in movement, here is not the Ātman which is said to be all pervasive and immutable but the subtle body of the sādhaka. During this journey along the path of Gods, the sādhakas are helped and directed by the deties assigned to different stations. After that a non-human guide (amānava puruṣaḥ) conducts

18. Ad. concept of Jīvanmukti : L. K. L. Srivāstava pp. 13-14.

them to Brahman (saguṇa Brahman). The image-worshippers, in the realm of Kārya Brahman, 'enjoy a state of quasi-deliverance consisting in hightened powers of enjoyment.' They do not return to the lower region of Saṁsāra any more. In due course of time they attain immortality.

Śaṅkara accepts this type of mukti because ultimately he has to rely upon Śruti and Śruti recognises such type of mukti. Though Advaita Vedānta subscribes faith to kramamukti, yet even here, unqualified deliverance can only be attained by the attainment of the right knowledge (Samyag darśana) of the Absolute. It may be 'Kramamukti' but it is attained by 'knowledge'.

Sadyomukti

Sadyo means immediate and mukti means deliverance. So Sadyomukti means immediate deliverance. This may also be taken as direct deliverance. It advocates that the aspirant passes to the final liberation at once with the dawn of the self-knowledge. In his commentary on Gīta, Śaṅkara says that Sadyomukti is attained by those aspirants who have established themselves well in the right vision (saṁyag-darśana) of the Absolute whom Śaṅkara refers here are, those whose happiness is within themselves, who delight within themselves, whose light is within, those who are yogin, and have become one with Brahman. Their sins have vanished away; their doubts have been cast away, they have perfect control over themselves and their mission of life is only to promote the welfare of all living beings. They have known the self and are,thus, delivered both in this life and hereafter. Some of the prominent post-Śāṅkarites lay much stress on the acceptance of Sadyomukti. According to them the only satisfactory form of mukti which can be obtained by knowledge of Brahman, should be Sadyomukti, that after the dawn of saving knowledge neither Avidyā nor its products e.g. body and the mind can even appear to exist.

(III) POST-ŚAṄKARA ADVAITINS

(Pre-Vidyāraṇya)

Realization of the identity between the self and the Brahman is liberation from bondage. This is a state of Mokṣa according to Advaitins but even after attaining the knowledge of Brahman and transcending all distinctions of the world by knowing it to be illusory, an individual may continue to have his bodily existence in the world. Such a liberated person is called a Jīvanmukta.

Śaṅkara admits the possiblity of Jīvanmukti. The concept of Jīvanmukti i.e. liberation while living has been accepted and has become welknown in later Advaitins. The Advaitic literature has been flourished after Śaṅkara by his pupils and by other Advaitins like Mandana Miśrā, Sureśvara, Vācapati Miśrā, Prakāśātman, Citsukhācārya, Ānandagiri, Vidyāraṇya and many others. Let us see the concept of Jīvanmukti in detail according to them separately.

As the potter's wheel continues to revolve though the stick that gives movement is detached, so the bodily existence continues though the Brahman knowledge is attained by the mumukṣu, because body is the product of karmas. Body does not immediately drop away even after the nescience is annihilated, by the knowledge of Brahman. He is called a Jīvanmukta. As soon as he realises that 'I am Brahman,' Jīvanmukta does not identify himself with his body. The world still appears before him but he is not deceived by it. He does not feel any desire for the world's objects. He is in the world and yet out of it. It is a state of perception attained here. To a Jīvanmukta world exists as a dream or mirage in a desert. He is steady and permanently fixed in his own nature. He is a living God on earth, immortal among mortals, perfect among imperfects, infinite among finites.[19]

---

19. Ātman and Mokṣa : G.N. Joshi, pp. 497-8

Maḍana Miśra

Manḍana supports the doctrine of Jīvanmukti. Manḍana (800 A.D.) is supposed to be contemporary to Śaṅkara. He accepts Jīvanmukti, but in certain features it differs from that of Śaṅkara. On the logical ground he does not accept Jīvanmukti Sadyomukti alone is logical for him. He thinks that at the dawn of knowledge, ignorance and all karmas, the fructified and unfructified disappear. Then the question arises that if all karmas are destroyed by knowledge, even those karmas which cause the existence of the present body, then mukta will be disembodied immediately. Here we come to know that Manḍana goes against Śaṅkara's treatment of Prārabdha Karma. On the basis of this assumption Manḍana takes a Jīvanmukta or a sthitaprajñā of the Gītā as still a sādhaka and not a siddha who has totally annihilated Avidyā for which a total annihilation of nescience the body must necessarily fall off.

But this view of Manḍana contradicts to that of Chāndogya (6.14.2) which says that the organs of a released soul do not depart and final release (videhamukti) is attained at the fall of the body. It also contradicts of the experiences of saints who have attained perfection in their life.

In order to save his position, Manḍana has to interpret his position in two ways—(i) that Brahman-realization brings about the total annihilition of all karmas; and (ii) in some cases even after realising Brahman, the body in which realization is achieved, does not fall off and persists for sometimes as a result of a trace of Avidyā. It persists in the form of prārabdha and that this condition is described as liberated in living state, i.e. jīvanmukti. The former of the two views, rules our Jīvanmukti while the later supports it. With the same in view, he has to accept two interpretations of the ideal of sthitaprajñā (Gītā 2.54). From the point of viewof Sadyomukti, he is a sādhaka,but from the point of view of jīvanmukti, he is a siddha. Manḍana believes that the direct realization of the Absolute unrelated Brahman

(निर्गुण ब्रह्मसाक्षात्कार) results from constant contemplation on the knowledge of the truth received from the Upaniṣadic texts. Karmas are also of as much service to the seeker after truth as a horse is useful to a traveller for reaching the destination shortly and safely. He also accepts meditation on 'AUM' as means to attain self-realization. Sanyāsa he thinks is the precondition of Brahman realization.[20]

Sureśvara (800 A.D.)

As far as the concept of Jīvanmukti is concerned, Sureśvara follows the view of Śaṅkara. In his Naiṣkarmyasiddhi he deals with the ideal of 'Jīvamukti' is detail. He opines that the body of the liberated person remains due to prārabdha karma even if he has attained the knowledge of the self. There is no conflict between the rise of the self-knowledge and subsistence of the prārabdha.He explains—just as an unrooted tree naturally perishes through withering, so the body of the enlightened man perishes through the consumption of his unexpanded prārabdha karma.[21] thus he accepts Jīvanmukti. He holds that jñāna and karma are means to attain Jīvanmukti and sanyāsa is indispensible for it.

Sarvajñātma Muni (900 A.D) :—The author of Śaṅkṣepa-śārīraka is supposed to be the successor of kāmakotipītha with Sureśvara—his preceptor and protector.

His views regarding the teachings of Advaita Vedāntā resemble with those of Śaṅkara and Sureśvara, and are in contrast with those of Maṇḍana. But in some respects, he joins hands with Maṇḍana and seems to deviate from what Śaṅkara and Sureśvara had said.

Sarvajñātman does not give full support to the concept of Jīvanmukti in the beginning. Like Maṇḍana, he also does not think the possiblity of the concept on logical ground because according to him, neither Avidyā nor its products such as body, can exist after the dawn of knowledge. It is all destroyed in its

20. Advaitic concept of Jīvanmukti : L.K.L. Shrivāstava, p. 141.
21. Ibid., p. 144.

fire. Thus the mukti, acceptable on this ground, is Sadyoumukti (immediate deliverance). To strengthen his view he says that all the references to Jīvanmukti are only eulogistic in their import. They encourage the aspirant by exalting the idea of deliverance or strive for it.

But in subsequent verses of his 'Sankṣepaśārīraka', he accepts that there is deliverance in embodied existence. He also opines that even after one has attained the knowledge of the self, there remains a trace of Avidyā which causes the present body of a mukta. But this trace according to him is not Avidyā or part of it. Such an acceptance will contradict liberation itself.

The trace of Avidyā is a mere shadow or apperance of Avidyā which causes the awareness of duality. Thus he accepts Jīvanmukti on the basis of experience. He says that experience is the only proof of the concept of Jīvanmukti. Thus he comes closer to the view of Śaṅkara. In this way Sarvajñātman accepts Jīvanmukti on the basis of experience, though not on logical ground.

Regarding the means to Jīvanmukti, Sarvajñātman gives primacy to knowledge arisen directly from the Upaniṣadic texts. Action, worship and yoga have secondary place in the scheme. He also accepts sanyāsa as a necessary precondition of attaining Jīvanmukti.[22]

Thus we come to know that there is difference of opinion between Maṇḍana Mīśra and Sureśvara regarding the concept of Jīvanmukti. The diffeence is based on the fact that Maṇḍana's expositionof the concept is based mainly on pre-Śaṅkara phase of it, while Sureśvara is exposition of that doctrine is entirely dependent upon its Śaṅkara-phase. The influence of these two Advaitins over their successors is evident from their division in two camps, namely Bhāmati school and Vivaraṇa school. Most of the distinctive features of Vācaspati school (Bhāmati school) have their roots in Maṇḍana views as set forth in the Brahmasiddhi and most of the distinctive features of the Vivaraṇa school are derived from Sureśvara's views as set forth

22. Ibid., p. 146-147.

in the Vārtikā and the Naiṣkarmyasiddhi. Let us observe the development of this concept in these two schools.

Bhāmati School—Vācaspati Miśrā

Vacaspati Miśra is the founder of the Bhāmati school (840A.D.). He is well-known for his commentary on Śaṅkara's Śārīraka-bhaṣya known as Bhāmati. He also wrote Tattva-samīkṣā, a commentary on Maṇḍana's Brahmasiddhi.[23]

Vācaspati follows Mandana Mishra in many respects but he has chosen to deviate from Maṇḍana and follow Śaṅkara in respect of Jīvanmukti. He speaks like Śaṅkara that the immedite experience of Brahman quickly destroys the accumulated (sancita) karma and future (Āgami) karmas, but can not destroy prārabdha karmas, which have become mature, active and fruitful. They are stronger than others which are dormant and inactive. He also accepts that a trace of Avidyā remains after the attachment of knowledge which causes the present existence of the 'mukta' as a momentum of a potter's wheel. He establishes the persistence of the trace of Avidyā on the illustration of the rope-snake also. As even after the illusion that the rope is not the snake the person in question, is not without the trace of fear. Similarly after the attachment of knowledge a trace of Avidyā remains.

Vācaspati Miśrā not only accepts the possibility of Jīvanmukti, but also proves it by giving some arguments. In his Bhāmati, he refutes, the contentions of the opponents that the body of the released soul can not continue due to the impression of Avidyā because Avidyā is not real. It is said that only a real entity can leave behind its impressions, but since merits and demerits (which cause the present body's existence) are constructed by Māyā which is unreal, so with the cessation of Avidyā merits and demerits are also detroyed. Hence, the presence of the mukta's body after the removal of Avidyā, is out of question. Since the impression of Avidyā is not real, it can not continue to exist.

23. A history of Indian Philosophy : S. N. Desgupta Vol. II., p. 106.

In reply to the above contention, Vācaspati Miśrā says that the nature of Avidyā is indescribable (anirvacanīya) and hence its effects are also indescribable. The production and desruction of merits and demerits are also of the same nature. Hence he concludes that this is not unreasonable to say that a trace of Avidyā remains even after the attainment of self-realization.

Vācaspati Miśrā also replies the contention of Maṇḍana Mishra that a person, whose wisdom is firm (sthitaprajñā), has not attained Brahman and that he is still an aspirant (sādhaka) after realization. He says that such a person has realised the self and is perfect. He has achieved embodied release.

According to Vācapati the performance of prescribed duties such as sacrifice, charity and penance is preparatory to the emergence of knowledge. It produces desire to know Brahman (viviḍiśā). Karmas, sense-control, mind-control are the subsidiary means to attain liberation. They purify the mind. Immediate knowledge of Brahman desires from the mind itself when it is purified. He believes that liberation in life is caused by the knowledge of Brahman. The means of the attainment of this knowledge are listening (śravaṇa), contemplating (manana) and nididhyāsana. Hearing a Vedic text generates undoubted knowledge of Brahman because the Vedas are the only (pramāṇa) proof in regard to it. Thus Vācaspati Miśrā accepts the ideal of Jīvanmukti on the line of Śaṅkara.[24]

Prakāśātman (Vivaraṇa school)

Prakāśātman is the founder of Vivaraṇa school. He accepts like his predecessors, that the immediate knowledge of Brahman can be attained, while one is in bodily existence. He also holds that the trace of Avidyā (Avidyāleṣa) in the form of prarabdha karma remains even after the attainment of self-knowledge. So according to him, there is no conflict between the rise of saving knowledge and the existence of the body. He establishes the ideal of Jīvanmukti by replying to contentions of the opponents

24. Ad. concept of Jīvanmukti : L. K. L. Shrivāstava, p. 148, 150

who hold the case otherwise. In replying to the objection that the experience of the identity of a jīva with Brahman and the experience of duality can not co-exist with each other after the destruction of Avidyā and karmas. Prakāśātman says that sometimes there is the experience of the identity of a soul with Brahman in super conscious trance and that some time there is the experience of duality due to the flaw of prārabdha karmas.[25] He takes the support of Śrutis in proving that the knowledge of Brahman has not the power of destroying prārabdha karmas which have already begun to bear fruit. The opponents content that a living person can not acquire the immediate knowledge of Brahman, because the knowledge of Brahman destroys all karmas including those of prarabdha ones. To this connection Prakāśātman replies that Śruti and Smriti assert that Vyāsa and others achieved immediate knowledge of Brahman in embodied existence due to prārabdha karmas and attained embodied liberation. A knowler of Brahman, he says, has a semblance of experience of duality, for his actions are not motivated by attachment, aversion and desire.

He argues that Avidyā is false knowledge and that it has an impression like odour, aquality, which has an impression. As knowledge has an impression, similarly a false knowledge has also an impression. The continued experience of Brahman gradually destroys the impression of Avidyā and when it completely destroys the impression, the body meets its decay and the Jīvanmukta attains the disembodied liberation (विदेहमुक्ति).[26] Thus we see that Prakāśātman establishes the genuinity of the concept of Jīvanmukti on the basis of reasoning and authority both.

Prakāśātman holds that the performance of rituals etc. are responsible for giving rise to knowledge of Brahman. He thinks that the results of the performed rituals are operative till the rise of the knowledge of Brahman. His view is known as

25. Pañcapādikāvivaraṇa, p. 786.
26. Ibid., p. 787.

'Vedanāpakṣa'. He holds karma, mind control, sense-control etc. as prerequisite to the realization of the self.

Prakāśātman believes that Upaniṣadic texts give rise to the immediate knowledge of Brahman. Śravaṇa, manan and nididhyāsana are the means to attain jīvanmukti, but Śravaṇa is the principle means where as manan and nididhyāsana are its auxilliaries.[27]

Thus we see that both the advocates of Bhāmati and the Vivarana school accept that even after the dawn of knowledge, a trace of Avidyā remains which causes the present body of a mukta. This trace of Avidyā thus makes Jīvanmukti possible. But they differ on the conception of the means to the attainment of Jīvanmukti.

Ānandānubhava, Citsukhācārya and Ānandagiri are thinkers of Vivaraṇa school. Let us see their opinion on this doctrine of Jīvanmukti very briefly.

Ānandānubhava (between 850 to 1220 A.D.)

Ānandānubhava accepts the concept of Jīvanmukti. He defines Jīvanmukti as a cessation of Avidyā. One whose Avidyā is eradicated is a Jīvanmukta.

He also accepts that a trace of Avidyā remains even after attainment of knowledge. It is due to this impression or trace of Avidyā, a perticular state of Avidyā that a mukta holds his body and remains embodied till his prārabdha karmas are exhausted. This is known as Jīvanmukti. But he says that even Jīvanmukti, from the ultimate point of view must be considered to be Māyā. Brahman or the self, which is non-dual, can never be said to be born or destroyed. In the abscence of creation and destruction there is no bondage and in the abscence of bondage there is no seeker after liberation and there is none free from bondage. Regarding renunciation, karma and jñāna, he holds the same view as propounded by Śaṅkara.

---

27. Ibid., p. 403, 410, 413.

Citsukhācārya (13th century)

He is one of the most important teachers of Advaita Vedānta. He contributes to the development of dialectical phase of Advaita philosophy of Śaṅkara. His most celebrated work is 'Tattva Pradipikā' or 'Citsukhi'. He holds that point of utmost importance in the spiritual of the Advaita Vedānta is that every seeker after mukti may hopefully aim at the goal of Jīvanmukti. Like other Advaitins he too, holds that after the attainment of Brahman-knowledge, Avidyā is destroyed but a trace of it remains. It is due to this trace its effects-prarabdha karma—also remains which holds the body alive. In order to explain this trace of Avidyā, Citsukha recognises two functions of Avidyā (i) concealing power (āvaraṇa śakti) and (ii) projecting functions. Only the concealing function of Avidyā is destroyed by knowledge. The projecting function-cause of apparent objects is destroyed when the prārabdha karma is destroyed. He too holds that a sthitaprajñā and a guṇātīta are not aspirant for release, but a perfect one. Citsukha holds that the concept of Jīvanmukti has its deep philosophical and scriptural import. He in the last sentene of Tattvapradīpikā strongly urges that the importance of Jīvanmukti can not be denied on the basis of mere prejudice against Advaitism. It is a fact of living experience and is datum to which history bears witness.[28]

Ānandagiri (13th century)

Ānandagiri also accepts Jīvanmukti. He argues in favour of it that a leśa of Avidyā remains even after the knowledge has arisen. He proves it by an analogy that even after the rise of the sun, trace of darkness remains, similarly after the rise of Brahman-knowlede a trace of Avidyā remains till the prarabdha-karmas are exhausted. This trace of Avidyā makes Jīvanmukti possible. Regarding the means of Jīvanmukti, he accepts the views of the Vivaraṇa school.

Thus we see that all the pre-Vidyāraṇya Advaitic thinkers

28. Ad. concept of Jīvanmukti : L. K. L. Shrivāstava, p. 154, 5, 6.
29. Ad. concept of Jīvanmukti : L. K. L. Shrivāstava, p. 154, 5, 6.

accept Jīvanmukti in someway or the other. Some of them try to discard its possibility on the logical ground but ultimately they rely on experience. Regarding the means of the attainment of Jīvanmukti, all the preceptors hold knowledge as to sole means to attain Jīvanmukti. Almost all the teachers before Vidyāraṇya accept the indispensibility of Sanyāsa for the attainment of Jīvanmukti. Vidyāraṇya not only supports Jīvanmukti, but he wrote a separate work namely 'Jīvanmuktiviveka' wherein he insists the importance of Jīvanmukti He elaborately discusses what Jīvanmukti means for him, what is videhamukti, the importance of Sanyāsa, yogic practices, at and the means to Jīvanmukti.

PART - II

Chapter - V

SHRI VIDYĀRAṆYASWĀMĪ

Among all the six orthodox systems, namely Nyāya-Vaiśeṣīka, Sāṁkhya-yoga, Pūrva Mimāṁsā-Uttara Mimāṁsā (Vedānta), Uttara Mimāṁsā claims to be the most representative of India's rich tradition in spiritual thought. Among all the six Vedāntic schools of thoght namely Advaita-Vedānta of Śaṅkara, Bhedābhedavāda of Bhāskara, Viśiṣṭādvaita of Rāmānuja, Dvaita Vedānta of Madhvācārya, Dvaitādvaita of Nimbārka, Śuddhādvaita of Vallabha, Śaṅkara's Advaita Vedānta is unanimously held to be the most important both as a system of metaphysics and as a doctrine of Mokṣa.

What makes the Vedānta both pertinent and interesting to modern students of religous philosophy or philosophic religion is its conviction of the inherent divinity of man which implied emphasis on the dignity of the human individual. The individual of the Vedāntic conception is potentially divine, what is more, he can realise his divinity here on this earth while still residing in the phenomenal order of existence.

The doctrine of 'jīvanmukti' maintains on the one hand that the aspirant can attain and enjoy the state of spiritual fulfilment here and now while embodied, on the other, it affirms faith in the capacity of man to work out his salvation by his own efforts. In as much as the Vedānta identifies the individual self with the Ultimate Reality, it proves for the possibility of the attainment of 'liberation in life' or 'jīvanmukti' without dependence on any agency external to man.[1]

---

1. Advaitic concept of Jīvanmukti : L.K.L. Shrivastava, pp. V, VI.

Life, Date and Works of Shri Vidyāraṇya

No clear and authentic account can be given of the life and works of Shri Vidyāraṇyaswāmī. Neither the exact birth date and the place of birth can be given. Whatever we can say are mere conjectures and are based on the tradition and the views of eminent scholars. He is said to have flourished towards the close of the 14th century A.D. His time has been fixed by the scholars to be A.D. 1297-1386 and this is in agreement with the tradition that he lived for 90 years.[2]

From his own works like 'Jaimininyāyamālāvistāra' etc. and from an inscription of the Arulal Perumal temple at Conjeevaram, it is known that Mādhava (Vīdyāraṇya) was a brahmin of Bhāradvāja gotra. His father's name was Māyaṇa and mother's name was Shrimati and that he had two brothers Sāyaṇa and Bhoganātha. In the beginning of his Parāśaramādhavīya he introduces himself thus :

श्रीमती जननी यस्य, सुकीर्तः मायणः पिता।
सायणो भोगनाथश्च महोबुध्दी सहोदरः।

बोधायनम् सूत्रम् यस्य
सखा यस्य च यजुसि।
भारव्दाज कुलम् यस्य
सर्वज्ञः स हि माधवः।[3]

He was born in a poor brahmin family near Hampi in the region of river Tungabhadrā. though brought up in poverty, all the three brothers became versatile scholars in all branches of learning. Bhoganāth took to the order of Sanyāsa in early life. Sāyaṇa and Mādhava were the authors of many works on religion and philosophy. (The name of Vidyāraṇya's before renunciation was Mādhavācārya). The famous commentary

2. History of Indian Philosophy : S.N. Dasgupta Vol. II, p. 419.
3. (Preface) Vivaraṇa Prameya Saṅgraha : Pandit Lalit Prasad Dabaral Hindu Tr. pp.1, 2.

of Ṛg-Veda, though a work of Sāyaṇa, was probably a combined work of theirs.[4] Apart from these two brothers, Mādhava had a sister by name Singale, whose son was Lakṣ midhara the author of 'Advaita Mākaranda', who after taking up asceticism was known as 'Kṛṣṇendrasaraswati'.

For relief from poverty, Mādhavācārya is said to have performed austerities at the shrine of Devi Bhuvaneśuari at Hampi, but Devi revealed to him that in that life he was not destined to be rich himself, but he would be able to help others to become rich. This was an indication of the great part he was to pay in the political life of his times.

In his fortieth year, he became associated with the founders of the Vijaynagara empire—Hari-Hara I and his brother Bukka I (In fact five brothers, sons of a chieftian named Saṅgama were responsible for building up empire. They were Harihara, Kampa, Bukka, Marappa and Mudappā, but Harihara-Bukka were mainly responsible.[6]) who began the consolidation of that state by 1336. He served under three successive kings as chief minister and built up the greatness and prosperity of that kingdom until he retired in about 1380 to take to the life of Sanyāsa at the age of 85. He became the head of Śringeri Math for a few years and passed away at the age of 91 in 1386.[7]

Mādhavācārya is said to have three gurus viz. Vidyātīrtha, Bhāratitīrtha and Śankarānanda.[8] Vidyātīrtha was considered by him as an incarnatin of Maheśvara. He was a great and unique personality, who occupied the Honourable teacher's post (guru-pith) of Advaita Monastery (Matha) at Śringeri at that time. Of his disciples, Vidyāraṇya was one of the foremost.

---

4. Śaṅkara Digvijaya : Tr. by Swami Tapasyānanda.
5. The Anubhutiprakāśa of Vidyāraṇya : (Godabarisha Mishra)
6. Ibid.
7. Śaṅkara Digvijaya : Tr. by Swāmi Tapasyānanda.
8. A history of Indian Philosophy : S.N. Dasgupta Vol. II p. 214 (footnote)

He was one of the eight disciples ordained by him for the eight Monasteries established by him (Vidyātīrtha). This is the ascetic dweller at Śringeri who is credited by the tradition with the foundation of Vijayangar Empire. Inscriptional evidences also show that he (Vidyātīrtha) was highly esteemed both as a temporal and spiritual guide not only of Mādhava but also of the famous king of Vijayanagara, Bukka I, whom Mādhava eulogies in his Nyāyamālavistāra and other works.[9]

So first guru of Mādhava was Vidyātīrtha. After his death, he might have studied under his disciple Bhāratitīrtha Vidyāraṅya. In his old age, it seems, he took initiation 'sanyāsa dikṣā' from Śaṇkarānanda and became a sanyāsin. He, then adorned the spiritual headship of Śringeri Matha and assumed the appellation Vidyāraṇya.[10]

As the term 'Vidyāraṇya' i.e. 'forest of learning' itself indicates that Mādhava was a versatile genuine. He is known to have written on almost every important branch of literaure, in his time with such a grasp and finish as would suprise the most accurate writer of the present day.[11]

In the field of religion and philosophy he has been credited with the greatest of the post Śaṅkara Advaitavādins. There is scarcely a branch of learning considered by Hindus as important to whcih he or the scholars who gathered round him, did not make valuable contribution and it is to his commentaries that the modern world owe its knowledge of the traditions of the sacred books. He was at the same time a philosopher, a grammarian, a poet, commentator, logician and a versatile scholar.

Tradition also affirms that Vidyāraṇya was excellent in Brahmavidyā. In the beginning of his 'Translation of Pañcadaśi, Shri Hari Prasad Shastri, eulogises, Vidyāraṇya as jīvanmukta.

---

9. Ad. Con. of J.M. : L.K.L. Shrivāstava, p. 32.
10. Idid : p. 33.
11. Intoduction to Jīvanmukti Viveka, p. XI

He says—

Salutation to the great Sage Vidyāraṇya, who has realised
the unity of Brahman,
whose learning is as fathomless
as the ocean,
And who is an infinite reservior
of peace and bliss.[12]

Works of Vidyāraṇya

Vidyāraṇya had contributed much to every branch of Sanskrit literature. Some of his works are :

Pūrvamimāmsā	:	Jaiminīyanyāyamālā
Astronomy and	:	Kālanirṇaya
Astrology	:	Mādhavīyajyotiṣa
	:	Laghujātakaṭīkā
	:	Sūryasiddhāntaṭīkā
Mantraśāstra	:	Shrividyārthadīpikā
Music	:	Saṅgitasāra
Advaitavedānta	:	Anubhūtiprakāśa
	:	Aparokṣānubhūtīkā
	:	Bṛhadāraṇyakā Vārtikāsāra
	:	Jīvanmuktiviveka
	:	Pañcadaśi
	:	Vivarañaprameyasaṅgraha
	:	Commentaries called Dīpikā

On—Aitareya, Chāndogya, Kaivalya, Nṛsiṁhottaratāpini and Taittaīrīya Upaniṣads.

Grammar	:	Mādhavīyadhātuvṛtti

Besides these—Sarvadarśanasangraha, Śaṅkara-digvijaya, Dṛk-Dṛṣya-Viveka, Mādhavīya-dhatuvṛtti, Parāśaramādhavīya Jaimininyāyamālāvistāra etc. have been assigned to the authorship of Shri Vidyāraṇya.[13]

12. Ad. con of J.M. : L.K.L. Shrivāstava, p. 33-34.
13. Anubhūtiprakāśa of Vidyāraṇya : Godābarisha Mishra (Introduction, p. 111.

The 'Jīvanmukti viveka' seems to be the last work written by Vidyāraṇya.[14]

Identity of Vidyāraṇya

(1) It is a controvertial issue whether Mādhavācārya is identical with Vidyāraṇya, but in fact these two are idential with the latter being only the surname of former.[15]

(2) Again Mādhavācārya and Mādhavamantrin are also different. Mādhavamantrin too was a minister under the Vijayanagar emperors. But this Mādhava was the son of Cauṇḍa of 'Angirasagotra' and was the disciple of Kāśivilāsa Kriyāśakti, a Śaivite teacher.[16]

There are many more controverses, but we come to the conclusion that though the identity of Vidyāraṇya, an ascetic, with the founder of Vijayanagar empire is a difficult problem to solve, yet on the basis of the views of scholars we may make a guess that the same Vidyāraṇya who adorned as honourable teacher ('gurupitha') of Śrinageri Math was the spiritual guide of the early kings of Vijaynagar Empire. with the advent of Vidyāraṇya, Śringeri became a state within the state, and its guru, a Rajguru with the royal position.[17]

Influences on the Writings of Vidyāraṇya

Vidyāraṇya represents the Advaitic school of thought. The influences of his predecessors are evident on his writings. Besides them, he is influenced by other works also. He seems to have derived his main inspiration from the Yoga Vāsiṣṭthā.[18] By his time Yoga-Vāsiṣṭha had become an authoritative work. It has been freely quoted in his writings. Dr. B.L. Atreya says that Vidyāraṇya bases his 'Jīvanmuktiviveka' mainly on Yoga-

---

14. Introduction to J.M.V. Eg. Tr., p. XI.
15. Ad. con. of Jīvanmukti L. K. L. Shrivāstava, p. 34.
16. Anubhūtiprakāśa of Vidyāraṇya : Godabarish Mishra., p. ii
17. Ad. Con. of J.M. : L.K.L. Srivastava, p. 33.
18. A history of Indian Philosophy Vol. II : S.N. Deshgupta, p. 252.

Vāsiṣṭha, quoting no less than 253 ślokas from it in support of his view.[19]

Besides Yoga-vāsiṣṭha, he draws upon the following Upaniṣads : Bṛhdāranyaka, Chāndogya, Kaṭha, Muṇdaka, Māndukya, Brahma, Paramahaṁsa, Jābāla, Śvetāśvatara, Āruṇi etc. He also quotes from Maitreyi Brāhmana, Taittirīya-Brāhmaṇa, Kauśitāki-Brāhmaṇa, Antaryāmi-Brāhmaṇa, Gītā, Bhāgavat, Bṛahaspati-Smṛiti, Yama-Smṛiti, Āpastamba-Smriti, Atri-smriti, Sūtta-Samhitā, Naiṣkarmya-siddhi, Pañcapādikā, Viṣṇu-Purāṇa, Yoga-sūtra, Vyāsa-Bhāṣya, Siddha-Gītā, Mahābhārat, Vālmiki Ramāyaṇa, Kāvaseya-Gītā, etc. Besides he mentions from the works of Yama, Parāśara, Bodhāyana, Medhātithi, etc. These influences of Śruti and smriti and the works of ancient scholars are very clearly visible in 'Jīvanmukti-viveka'. His other works are also influenced in general. His such clinging to the tradition makes us sometime sceptical of his originality. But his way of presenting concept of Advaita Vedānta in a clear and lucid manner, is really marvellous. Herein lies his originality.

Let us take a brief review of his Jīvanmuktiviveka and some other important works.

Jīvanmuktiviveka

'Jīvanmuktiviveka' or 'Path to liberation in life' is a well-known and unique work by Shri Vidyāraṇya. The path to liberation, according to him is indeed in one word-renunciation. Though the word has received a number of different shades of meaning from several writers, Vidyāraṇya means it in formal orthodox sense in which the Hindu seers principally employed it. It is sanyāsa. Sanyāsa means not to have concern with the world neither bodily nor mentally,neither directly nor indirectly, to live in entire isolation, and wear the orthodox insignia of the order. Vidyāraṇya himself was a true brāhamaṇa and his very life nobly illustrates the truth of his teachings. He wrote this book after he renounced all the concern with the world i.e. he

19. The Philosophy of the Yoga-Vāsiṣṭha : B.L. Atreya, p. 12-13.

become a saṅyāsin. His life, spent in the midest of varied activities at the court of Bukka Rāya, had its culmination in the quiet bliss of supreme spiritual exaltation.[20]

The first chapter is Pramāṇaprakaraṇa. It oepns with the discussion on the nature of renunciation and the two divisions (vividiṣā and vidvat) are described and justified by scriptural texts and ancient examples. Then he explains nature of Jīvanmukti i.e. liberation in life. It is freedom from bondage. The discussion of the nature of Jīvanmukti leads to the description of the nature of videhamukti, i.e. liberation after death. The chapter ends with illustrative quotations from several spiritual texts to bear out the author in his arguments on the nature of these two kinds of liberation.[21]

The second chapter deals with the means to acquire the condition of 'Jīvanmukti'. These are three in number. They are knowledge, oblitaration of latent desire (vāsanā) and dissolution of mind. The whole question of liberation is easily solved, if we once acquire a true perspective of the nature of vāsanā and clearly understand the means to destroy it. This chapter gives a masterly exposition of this most important question in a thoroughly practical manner and the earnest student is sure to find here what he may elsewhere seek in vain. The aim and object of all methods of spritual exaltation is the life of the higher self, which is the ideal they invariably point to. Purification and magnanimity of the heart are inculcated as the most important practical steps towards its realization. He advices, 'conquer these latent desires and Jīvanmukti is within easy reach. Act without being mastered by the action and find this place of peace beyond the pale of the disturbing influence of mind and desire. He says "giving up all attachments from within one should acquire that limitless expansion of the heart which is the royal road to 'Jīvanmukti'.

The other path to liberation is the dissolution of the mind.

20. 'Jīvanmuktiviveka' of Vidyāraṇya (Liberation in Life): Eng. Tr. by Pandit s. Subramanya Shastri and T.R. Shrinivasa Ayyangar (Intic)
21. Ibid. : (Intro.).

The third chapter fully deads with this subject. The mind is made up of latent desires, act and react upon each other and obliteration of vāsanā is never confirmed without the dissolution of the mind. The two methods of dissolving the mind, the physical and the mental are described. Then are discussed the obstacles in the way of ecstatic trance, the true condition of the liberated life.[22]

The fourth chapter aptly opens with the question of what is the aim of or use of 'Jīvanmukti'. The answer is five-fold. (i) Preservation of knowledge (ii) practice of Penance (iii) Universal Love (iv) Destruction of Pain and Misery (v) Experience of Supreme Bliss. These are severally described and discussed in the chapter.[23]

The fifth chapter concludes the original inquiry into the nature of renunciation the true path. It is the renunciation of the Enlightened. Renunciation which is a necessary condition preceding the realization of jīvanmukti, ripens into that true, renunciation which is known as the condition of the Paramahaṃsa. Parama means great and haṃsa as interpreted by Shri Śaṅkara, means the destroyer of all Avidyā, and hence Hamsa is the Ātman, the self. Whence Paramahaṃsa is the great self. Great being, the Being that is the All.24

Anubhūtiprakāśa

Anubhūtiprakāśa is a material exposition of the philosophy of twelve upaniṣads namely Aitareya, Taittirīya, Chāndogya, Muṇḍaka, Praśna, Kauṣitāki, Maitrāyani, Kaṭha, Śvetāśvatara, Bṛhadāraṇyaka, Kena, and Nṛsihottaratāpani.

Vidyāraṇya in the text 'Anubhūtiprakāśa' has expounded in a systematic manner the central philosophy of the Upaniṣads. He has shown that the Ultimate Reality is Brahman which is eternal, pure consciousness and non-dual. It is identical with the essential nature of the individual soul. And this is the logical

22. Ibid., p.
23. Ibid., p.
24. Jīvanmuktiviveka of Vidyāraṇya : Eng. Tr. by Pandit S. Shastri and T.R.S. Ayyangar.

significance of the major texts of the Upaniṣads such as 'Tat tvam Asi'.

Brahman which is non-dual is said in the Upaniṣads to be the cause of the world. It is owing to Māyā that it becomes the cause of the world. Vidyāraṇya makes it clear by saying that Māyā is the transformative material cause of the world and Brahman by being the substratum of Māyā whose transformation the world is, is viewed as the transfigurative material cause of the world.

Author emphasizes the fact that liberation being identical with Brahman is ever attained, yet through a mistaken notion of its not being attained, the aspirant longs for it and attains it (as if it were unattained) through the removal of Avidyā Attainment of liberation is thus figurative.

One who has realised the truth and who continues to exist in the body is a Jīvanmukta. He will be a mere witness of the happiness and misery which are the modes of mind, that are born out of the fructified merits and demerits. He will perceive the world as illusory. When the fructified merits and demerits are exhausted by providing happiness or misery which are only the modes of mind, the knower of the truth will be dissociated from the pycho-physical organism and would remain as Brahman. This is 'videhamukti.'[25]

Śaṅkara Digvijaya

The text is a master-piece of literature and philosophy, which the great mind like Vidyāraṇya had produced. It is a historical poem. He has given in this poem a precise summery of the ancient literature connected with the glorious life of Ācārya Śaṅkara. He says, 'In a small mirror, even a the big forehead of an elephant can be clearly seen. Even so, in this, my brief work, you will get the essence of all the extensive accounts about Śaṅkara and his teachings. Nothing is impossible for one to achieve, if backed by the grace of teacher (Guru) Śaṅkara was his spiritual Guru.

25. Anubhūtiprakāśa of Vidyāraṇya : (Godabarish Mishra Introduction).

In praise of Śaṅkara as a sanyāsin, Vidyāraṇya says, 'Śaṅkara, the greatest of ascetics, is a real Paramahaṁsa, a supreme swān, just as that noble bird is able to separate milk from water in spite of their indivisible mergence in each other, so has he been able to do the very difficult task of separating Brahman from the world of Saṁsāra which seems to have covered Him and hidden His identity. Thus revealing the latent spirit in the minds of men, he has helped to eliminate from them the polluted waters with which the life of samsāra has adulterated their spiritual identity. All knowledge in the relative world is false, it being an in extricable mixture of the true and the untrue like that of milk and water, caused by superimposition arising from ignorance. Even the wisest of men are, therefore, puzzled in the quest for truth. But the great swan Śaṅkara has been able to extract the milk of truth from the water of untruth and present the same for the benefit of mankind. Those who will not accept it are verily mean creatures like crows which stick to their habbit of eating the bitter margosa fruits even when better things are available."

"It is indeed very appropriate that Śaṅkara shares the name of Haṁsa (Parama-haṁsa) with the sun (Haṁsa). As the sun removes the darkness of night, Śaṅkara eliminates the darkness of ignorance from the minds of men. He is again a 'Mitra', a friend, of all like mitra, the sun, giving light to the whole world. As the sun gives the joy to the Nalikas (lotuses) by its light, Śaṅkara gives joy to the wise with Nalikas (pure truth with all untruth eliminated). As the sun by its light removes the sufferings of separation of Chakras (Chakrawāka birds), he removes the accumulated miseries of a wide circle of devotees (Suhrit-chakra). As the sun reveals all Arthas or objects of the world, Śaṅkara reveals the Truth (Artha) of the spirit of mankind.[26]

---

26. Śaṅkara—digvijaya by Mādhava Vīdyāraṇya Tr. by Swami Tapasyānanda.

Thus the greatness of Ācārya Śaṅkara is narrated in this work.

Pañcadaśi

It is a text, written by Vidyāraṇya, having fifteen chapters. It is 'Vedānta Pañcadaśi :

पञ्चाधिकानां दशानां प्रकरणानां समाहारः।

(A collection of ten plus five chapters as a numerical determinative compound).

The word 'Pañcadaśi' also means either the new moon or the full moon. So 'Pañcadaśi' is considered to mean moonlight. It is well known that moonlight is soothing and comforting in the night to those who are scorched by the heat of the sun during the day. Similarly the Seekers of Mokṣa (mumukṣus) who plung into Vedānta to discover the truth, in the beginning are faced with an uphill task of ascertaining the exact meaning which saps them away of their entire enthusiasm. They are no better than those who are parched in the scorching heat of summar. To such mumukṣus who are caught up in the heat of study of Vedānta, this text 'Pañcadaśi' serves as the moonlight, soothing them by an easily understandable mode of presentation of teaching to unfold the final purport.

Pañcadaśi contains (1) Viveka-Pañcaka, (2) Dipa-Pañcaka, (3) Ānanda-Pañcaka. The greatness of the text is that though all the chapters are interconnected, each by itself presents an independent mode of inquiry which is self-sufficient in ascertaing the truth.

Pañcadaśi unfolds the truth contained in the Upaniṣads.[27] Vidyāraṇya describes Brahman thus—Supreme Reality. Brahman is unaffected by names and forms. Just as rocks that lie burried in the river-bed are not affected by the flow of the streams, even so the imutable Brahman remains unchanged in spite of the fleeting thing of the world. The appearance of the world in Brahman is like the reflection of the heavens in the

27. Vedāntā Pañcadaśi : Com.by Swāmi Śuddhabodhānanda.

mirror. Without lookng at the mirror, it is not possible to behold the reflection of the sky. Similarly without the thought of Brahman, there is not even the thought about names and forms Superficial cognition of Brahman is vouchsafed even by the cognition of objects, But if that cognition is to be constant, we must cease from cognizing the world as real. When Brahman which is of the nature of truth, intelligence, bliss is seen, the cognition of names and forms vanishes of its own accord.[28]

—Pañcadaśi is one of the most favoured books which has found a permanent place in a study of the Advaita system.

Vivaraṇa-Prameya-Sangraha

Vivaraṇa-Prameya-Sangraha is important for Advaita dialectics. It is a summary of the topics dealt with the Vivaraṇa of the Prakāśātman, which is a gloss on the Padmapāda's Pañcapādikā. Pañcapādikā in its turn is a commentary on the Śānkara-bhāṣya on the first four aphorisms of the Vedānta-sūtra.

Brahman is described here as pure, flowless and undefiled. In reaching a place, a village not attained before, there requires action in the form of movement, in removing a desease, there requires action in the form of taking medicine etc.. But in the case of Brahman which is eternally attained, it needs only the removal of ignorance by knowledge. Knowledge is the path to perfection. Karmas are the cause of ignorance. The fire of knowledge reduces all karmas to ashes.

Vidyāraṇya emphasises the importance of renunciation here. He believes that renunciation is a potent weapon which can be wielded only by the strong and the skilled. When one realises that attachment to the things of the world is the source of misery and turns away from them one becomes eligible for renunciation. The day on which he is non-attached, that very day he is to renounce.

Śaṅkara lays down a four-fold constitution for eligibility

28. Pañcadaśi XIII 93-102.

viz.—(i) The discrimination of things eternal from non-eternal (ii) Non-attachment to the enjoyment of fruit here or hereafter (iii) The possession in abandance of means like calmness and equanimity and (iv) the desire for release. All these are means to release. He who is fully qualified by possessing the four fold requisite is to study the Vedānta texts under the guidance of Guru who is learned in the sacred lore (Śrotriya), well-established in Brahman (Brahma-niṣṭha). Scriptures enjoins for him who is eligible hearing (Śravaṇa) as principal, with reflection (manana) and contemplation (nididyāsana) as subsidiaries. Verbal testominy also is capable of generating the immediate knowledge of the self-luminious Brahman. It is directly the Karaṇa (instrument) of Brahman-intuition.

Advaitic view of Jīvanmukti is seen in Vivaraṇa-Prameya-Sangraha. A Jīvanmukta is he who is released even while being embodied. There is for him the persistence of the body until the fruition of the residue of prārabdha karma. Though knowledge of the truth destroys Avidyā that is the material cause of all karma without exception still it does not remove prārabdha karma since it is itself the fruit of that karma. Both the release and the bound have alike to reap the consequences of their prārabdha. Both have to walk in the thorny path, but while the enlightened with shoes on, the ignorant have no such protection to enable them to pass unhurt.

Dṛg-Dṛśya-Viveka

As the name indicates, it is an inquiry into the distinction of the 'seer' (Dṛg) and the 'seen' (Dṛśya)—an inquiry which is of the utmost importance for the understanding of the higher Vedānta Philosophy. The other name by which this treatise is known is Vākya-sudhā ascribed to Śankarācārya, which is also the name of commentary on it. This work has been acknowledged as a 'Prakaraṇa' treatise of the Vedānta Philosophy i.e. a book which though confirned to a perticular aspect of the subject matter, explains its chief purpose viz. the identification of jīva and Brahman, by following a perticular line of argument. The special feature of this book is its detailed description of the various kinds of Samādhi (concentration), the importance of which is always acknowledged by the students of Vedānta.

This work being very short, it necessarily compresses much information into a small space. The central theme of this work is that Brahman is realised through mystic or Yogic Samādhi (concentration). After this state is attained, one can see Brahman wherever one turns one's eyes.—With the disappearance of the attachment to the body and with the realisation of the Supreme Self, to whatever object the mind is directed one experiences Samādhi—The mind is completely absorbed in the contemplation of Brahman, the practitioner attains the Nirvikalpa-samādhi by merging the entire illusory phenomena in Brahman and by being indifferent to the manifested manifold (दृश्य) and such idea as indivisible (अखड) of the same nature (एकरस) etc. concentration becomes steady like the unflickering flame of a candle in a place free from wind. This samādhi should be practised uninterruptedly for a long time. Then only can the practitioner be firmly established in supreme knowledge. It beomes firmly grounded by long, constant effort with great love, for the end to be attained.

Chapter - VIII

JIVANMUKTIVIVEKA'

By Shri Vidyāraṇyaswāmī

TRADITIONAL PROOFS IN SUPPORT OF THE POSSIBILITY OF JĪVANMUKTI

Shri Vidyāraṇya Swamī is an outstanding exponent of the Advaitic religious thought epitomised in the concept of 'Jīvanmukti'. The views of Śaṅkara and of Vidyāraṇya on Jīvanmukti are nearly the same, but Vidyāraṇya has added many new details in the discussion of the problem. Though Śaṅkara has accepted and given, the explanation of Jīvanmukti, Vidyāranya has, more systematic account of the ideal of Jīvanmukti.

His 'Jīvanmuktiviveka' contains an exhaustive discussion of the subject. 'Jīvanmuktiviveka' is especially interesting because of the author's attempt to weave together a number of strands of Indian thought to make 'Yogic Advaita' While primarily an adherant of Advaita Vedānta, Vidyāraṇya also emphatically enjoins following the ascetic path, and refers far more often and more favourably' than does Śaṅkara to the Yogic practices of 'Patañjali' and 'Bhagavadgitā'. Another indication of Vidyāraṇya's 'Yogic Advaita' is his repeated citation of the 'Laghu Yoga Vasiṣṭha' at times the Jīvanmuktiviveka virtually becomes a commentary on Laghu Yoga Vāsiṭha. Deeply influenced by Laghu Yoga Vāsiṭha, the Jīvanmuktiviveka is more respectful of Vedic and Sanyāsa traditions linked with mainstream Advaita, 'while the Laghu Yoga Vāsiṭha has an even greater compliment of Saṁkhya, Paurāṇika and Buddhist ideas.[1]

The conception of 'Jīvanmukti' is based on self-certifying experience of the saints and mystics. Vidyāraṇya himself being a saint, a sanyasi and a Jīvanmukta. advocates the ideal of Jīvanmukti, proves its possibilty by quoting evidences from Srutis, Smritis and many other religious works. 'Jīvanmukta' is denoted as a guṇātīta. a sthitaprajña, a brāhmaṇa, a bhakta, an atīvarṇaśrami etc. in different works.

So, to prove the possiblity of Jīvanmukti, Vidyāraṇva quotes precious verses from ancient religious works and comments them carefully.

'Jīvanmuktiviveka' has been divided into five chapters: The first one is Pramāṇaprakaraṇa. i.e. Authority bearing testimony to Jīvanmukti. The second and third chapters deal with the means to attain Jīvanmukti. The means are knowledge, obliteration of latent impressions and dissolution of mind. The fourth chapter discusses the purpose of attainment of Jīvanmukti and last and the fifth chapter is on renunciation of the Knower or Vidvat Saṅyāsa prakaraṇa. Let us discuss every chapter thoroughly.

The first 'Pramāṇa Prakaraṇa' deals with the authority bearing testimony to Jīvanmukti. Here he discusses Vividisā"-Saṅyāśa, Vidvat-Saṅyāsa and the charactaristics of Jīvanmukti. Let us see Pramāṇa-Prakaraṇa in detail.

The chapter begins with venerating his Guru Vidyātīrtha and procedes ahead to explain Vividiṣā Sanyāsa and Vidvat Sanṅyasa. He says that Vividisā Sanyāsa is the cause of 'liberation after the dissolution of the body (i.e. Videha), and the Vidvat Sanyāsa is the cause of 'liberation in life' (i.e. Jīvan-mukti). Renunciation arises due to detachment and as soon as the detachment arises, one must renounce the world immedia-

---

1 Liberation While Living in the J.M.V. : Vidyāraṇya's Yogic Advaita By Andrew O. Fort Texas Christian University

tely This detachment is of two types sharp and sharper. Sharp detachment leads to the renunciation appropriate to the condition of Kuticaka. whereas sharper detachment makes a saṅyāśin, a haṁsa Kuticaka condition, when ripened develops into Bahūdaka whereas Haṁsa condition ripens into a Paramahaṁsa. the real path to direct self realization.[2]

Dull-detachment is the form of disgust generated, for the time-being, for the world and its goods, occassioned by the loss of a child, wife or wealth.

Sharp-detachment is the firm resolve of the intellect not to have a child, wife or wealth, in this life.

Sharper-detachment is the strong disgust in the form 'For me the whole of this world, whirling through the cycle of rebirths, shall never be' In dull detachment no renunciation is possible.

In the sharper kind of detachment two kinds of renunciation is possible A condition that of a Haṁsa which leads to Brahmaloka and the other that of a Paramahaṁsa which leads to final and absolute liberation.[3]

(i) Vidyāraṇya. then explains the Vividiṣā-Saṅyāsa or the renunciation of the seeker which is the cause of Videhamukti. Giving the reference of Bṛhadāraṅyaka. up he explains that Sanyāsins renounce the world for the purpose of self realization or Ātmānubhava: In smritis also it is said that:

ब्रह्मविज्ञान लाभाय परहंस समाव्हयः।
शान्तिदान्त्यादिभिः सर्वैः साधनैः सहितो भवेत्।।

(The one styled a Paramahamsa should, for the attainment of Brahmajñāna, have in him all the requisite means namely peace of mind, control of body and the like). As this saṅyāsa comes about by the strong desire for knowledge generated in the proper manner, in consequence of the study of the Veda,

2. Jīvanmuktiviveka p. 1.5
3. Ibid., p. 1/6-10

etc. (i.e. performance of religious duties certain austerities and the like, enjoined by the Veda), either in this or in some previous incarnation, it is called the renunciation of the seeker, i.e. Vividisa-Saṅyāsa, thinks Vidyāraṇya.

As written in Taittirīya and Kaivalyopaniṣad (3) : Some have attained immortality not by acts, nor by offspring, nor by wealth, but by renunciation alone. The Chaturdhāri tīka of Mokṣa-dharma says that females also are entitled to this kind of renunciation. It uses the word 'Bhikṣuki' for female mendicant. But such female mendicants may renounce before marriage or after the death of their husbands and may go about as religious mendicants may lean and hear the sutras dealing with Moksa (namely the Upanisads and Cognate literature), may meditate upon the Ātman [4]

Even the Gruhasthas and the Vānaprasthas, though they are unable, for some cause or other, to go into saṅyāsa they can attain mental renunciation for the purpose of attaining knowledge even while performing the duties peculiar to their respective spheres of life. Many instances of such knower of truth occur in the Śrutis, Smṛitis Purāṇas and ltihasas. as also in the world.[5]

Thus definied Vivīdisa-Sanyāsa, its charactaristics and who can attain it, by giving the proofs from Bṛhadāraṇyaka upaniṣad. Taittirīya, Āpastaṁba-dharmasūtra (Prasṇa Upaniṣad 2 - Khanda 22).

(ii) Vidvat-Saṅyāsa Renunciation by those who have realised the Supreme Truth, by properly hearning out study, reflection and assimilation that is Sravaṇa, manana and Nididhyāsana, is called the renunciation of the knower Sage-Yājñāvalkya is a good example of Vidvat Saṅyāsa. Vidyāraṇya mentions his example and says, 'The worshipful Yājñāvalkya. the crest-jewel of the learned, having discomfited in debate Āśvalāyana and other vipras, by variously demonstrating the Truth and having awakened Janaka, in variety of ways, by

4. Ibid p.4
5. Ibid p.4

disquisitions long and short, to the condition of Supreme detachment Vītarāgatva, set himself to enlighten Maitreyi, his wife, into the Truth and put to her saṅyāsa, as the thing he had immediately in view for himself. Having thus enlightened her. he became a Saṅyāsi'. All this is mentioned in the beginning of 'Maitreyi Brāhmaṇa. In Kaholabrāhmaṇa also we find this Saṅyāsa mentioned thus: Having realised that Ātman aforesaid, Brāhmanas renounce all desire for offspring, wealth and the world i.e. Putraiṣaṇā, Vittaiṣaṇā, and Lokeṣaṇā and go about as mendicant.[6]

Giving the reference of Śārira-Brahmana (4.4.22). Vidyāraṇya explains the point more clearly He says: Two kinds of renunciation Vidvat and Vividisā-Saṅyāsa are clearly mentioned in the Śārira Brāhmana. knowing it (i.e the Supreme Self), they melt in silence, desirous of that sphere (i.e. Ātmaloka), they enter Sanyāsa. The 'melting in silence' refers to the habit of contemplation & reflection and this is possible, only when there is nothing to distract the mind from it. whence, nothing but Sanyāsa is plainly implied in these words. The text further says : This is the reason why knowers of you have not wished for progeny, having as their plea - 'of what avail is progeny to us to whom this world, this Ātman is the Supreme Place ?' - They renounce all desire for offspring, wealth and world and betake themselves to alms.[7] The words 'this world', 'this Ātman' implies that the said world has been directly realised.

Both these kinds of Saṅyāsas are included in the order of Paramahaṁsa. But there is a degree of difference. The two renunciations have peculiarities of their own running counter to the real nature of each other.

In Jābāla-Śruti sage Yajñāvalkva says that Ātmajñāna (self-knowledge) is the real sacred thread of the seeker, The seeker is included in the denomination Paramahaṁsa, as he should not wear the sacred thread.

---

6. Br. UP. 3.5.1
7. J.M.V., p. 5

Regarding the Vividiṣā-Saṅyāsin, it further says - the Paramahaṁsa is one, bearing no marks, bound to no forms and behaving like mad man, though not mad. That he should throw oil in water, after reciting the words, "bhuḥ Swāhā" (be this offered to the bhūrloka), the three daṇḍas, the water-pot, the bag for holding alms, the cup, the filter-cloth, the tuft of hair on the head, the sacred thread and all, and then seek his own self (6.1-2).[8]

Regarding Vidvat-Saṅyāsa, which is the end and aim is sought thereafter, is thus described:- He is verily the Paramahaṁsa. who resumes the condition, he was in, at the moment of birth, frees himself from the pairs of opposites, neither receives nor possesses anything, ever walks in the exalted way of Brahman, with mind ever pure, goes about begging at large at the prescribed time, using his belly alone as his begging bowl; remaining indifferent to gain or loss, abides in any place, be it a deserted dwelling, temple, hay- rick, anthill, root of a tree, potters work-room, the house where the sacred fires are kept, the sand bank of a river, mountain cave or cleft, the hollow of a tree, rivulet or bare ground; avoids all exertion, is self less, is ever absorbed in divine contemplation, abides in the Ātman, ever bent on uprooting all actions, good, bad or indifferent, and shuffles off his mortal coil with the true spirit of renunciation (Jābāla up. 6-3). From this it is established that these two types of Sanyāsins are included in the order of Paramahaṁsa.

These two types of renunciations have peculiarities of their own. There is degrees of difference in them. In the Paramahaṁsa Up. when pupil Āruni asks to his teacher - O Lord! How can I completely abandon action? He asks here about the renunciation of the seeker which consists in the complete abandonment of the tuft of hair on the head, the sacred thread, the study of the Vedas, repetition of the Gāyatri and other actions of the kind. The teacher, Prajāpati lays down the acceptance of

8. J.M.V., p. 7

Bamboo-stick, a blanket. and a lion-cloth and says 'He should wash himself at each of the three saṅdhyās, be centred in the self at the moment of concentration and should constantly study that part of Vedas known as the Āraṇyakās, namely Upaniṣads.

Lord Prajapati describes the renunciation of the knower and at the end of his discussion he says: - The Paramahaṁsa has all his desires supremely satisfied, on his finally realising in himself that Brahman which is all Bliss and Light, ever full and thus experiencing the full force and meaning of the text 'Ahaṁ Brahmāsmi' (Paramahaṁsa up. 4)

Vidyāraṇya quotes precious verses one after another from the sacred texts and comments on them skillfully to emphasizes the importance of two types of Saṇyāsas.

Chapter - VII

NATURE OF JĪVANMUKTI AND VIDEHAMUKTI

As regards the nature of Jīvanmukti, Vidyāraṇya follows the view of Yoga-Vāsiṣṭha, though he supports it by many other scriptural references.[1] His originality lies in propounding the concept. He defines Jīvanmukti as freedom from bondage. He says :

जीवतः पुरुषस्य कर्तृत्वं भोक्तृत्वं सुखदुःखादि लक्षणश्चित्तधर्मः।
क्लेशरुपत्वाब्दन्धो भवति, तस्य निवारणं जीवन्मुक्तिः॥

Bondage of a living being, consists in those functions of the mind, which are characterized by feeling of pleasure and pain concomitent with action and enjoyment and which, therefore, are so many distractions (from the natural condition of bliss); freedom from this bondage is liberation in life.

Vidyāraṇya makes it clear that these bonds are attributed to the 'Mind' which, falsely identifies the self with the body, sense-organs etc., the resulting experience is characterized by pleasure and pain. This freedom is neither bodily nor mental. It is a freedom of a spirit. The freedom which is characterized by the removal of all bonds are external to the self. Had they been intrinsic to the self, they would not have been eradicated Again these bonds are not eradicated but are dissolved of itself

1. A History of Indian Philosophy : S.N. Dasgupta, Vol. II, p. 252

by knowledge. As we can not remove fluidity from the water, heat from fire, complete annihilation of the nature of a thing is impossible It's neutralisation is certainly possible. As the fluidity of water is counteracted by mixing earth with it. or as the heat of fire is neautralised by a (special) jewel or by some powerful incantation etc. so also all the modifications of the mind are neautralised by the practice of yoga [2]

Vidyāraṇya then explains how Jīvanmukti is attained. He says: One may concede that operative action is superior to so 'yoga' is superior to operative action. Such yoga is difficult to practice in case of short lived mortals like us, but there can hardly be any difficulty in bringing under control the active modifications of the mind, such as desire etc We must admit the potency of Sāstrīc efforts in this case. If we do not admit the same, then all the authoritative disciplines beginning with medicine and ending with liberation, will become utterly useless. Efforts may fail sometimes to its results, but it does not mean that one may give up them. Here Anandabodhācārya's words are quoted that —We do not give up food, because of the fear of indigestion. We do not refrain from preparing our dishes, because beggers are likely to swarm around.

Potency of Śāstric efforts is emphasized by giving the sixteen verses from Yoga Vāsistha. that reveal through the conversation between sage Vasiṣṭha and Rāma.

Everything (means progeny, wealth, heaven and Brahmaloka etc.) is invariably achieved by everyone, by well directed human efforts. Human efforts mean action of the agent consisting of such acts as Putrakamesti, husbandry, commerce, the jyotiṣ-toma, meditation on Brahman etc. Human initivative is of two kinds : in accord with the Śāstra and not in accord with the Śāstra. The first culminates in supreme good, the second in evil. Supreme good here means liberation.

Vasiṣṭha says. "If you are carried away by the force of

2. J.M.V., p.10

good impressions then you snail, by that very course, attain the eternal state in no length of time." 'Impressions' here denote the potentiality for action in the form of dharma and adharma i.e. religious merit and demerits stored up in the jīva.

"But if the force of bad impressions lead you to misery, then you must try to subdue it with your own efforts."

The words 'with your own efforts' are important, 'effort' here implies the observance of religious rites prescribed by the Śāstras, as capable of counteracting the force of bad impressions This force must be subdued by one's own efforts and not through other men. The river of impressions, which flows through good as well as bad courses, should be directed along the right (good) track by one's own efforts. Vidyāraṇya makes it clear that in the case of the alternative involving both kinds of impressions, though no effort is necessary in the case of good impressions, the bad ought to be displaced by the good, through such efforts as the Śāstras prescribe.

The mind which is engrossed in bad impressions, must be diverted into the right track, by strong personal efforts. 'Bad impressions' connote desire for adultery,theft and the like. The 'right track' consists in meditation on the import of the sacred texts, on divinities and the like.— Just as a child is prevented from eating clay and made to prefer eating fruits, instead, or as it is prevented from catching at pearls and jewels and taught to amuse itself with toys such as balls etc., so indeed, is it possible for the mind to be turned away by the company of the good, from those objects which are not conducive to its real happiness. Vidyārāṇya also compares mind with a restive animal. As restive animal is controlled by offering fresh grass, scratching its body etc. or it can be controlled by giving hard blows with a stick, harsh shouting etc. By the first method the animal can be controlled quickly in the same way mind can be controlled easily by evenness of feelings, looking with equal eye on foes and friends. By personal efforts consisting of prāṇāyāma pratyāhāra etc. (i.e. through Rājayoga), mind cannot be easily subdued, it takes time, but it can be controlled by'degrees. With the help of the softer yoga fruit can be attained in a short time. Good impressions begin speedily to arise in mind. While good impressions are being cultivated, the doubt may arise, whether such practice is complete or not. Even then the practice should surely be kept up. As long as you have not acquired complete mastery over the mind and have not realised the supreme state of realization go on following what is prescribed by the teachers, sacred books and the criteria (perception etc.). After that all latent desires will vanish, even the collection of good impressions also are given up without any pang. Persuing the very good goal sought by the wise, with sincere feeling and clear understanding, one realizes Brahman. Brahman which is ever blissful. This state is called Jīvanmukti. Such is the nature of Jīvanmukti explained by Vidyāraṇya.[3]

Characteristics of Jīvanmukti :

Giving the quotations from Lagu Yoga Vasiṣtha and Bhagawad gīta, Vidyāraṇya expresses his views on characteristics of Jīvanmukti and to whom Jīvanmukta can be called.

He writes. "One altogether Liberated is all the more so liberated" (Kaṭha Up. 5-1). It means one who is altogether liberated from the strong bonds of desires etc. is all the more so liberated on the fall of his body, from all possibility of future bonds of any kind.When all the desires occupying the heart fall off entirely, then indeed, does the mortal becomes immortal, and realises Brahman even here.[4]

Vidyāraṇya quotes both the charactaristics of a Jīvanmukta as well as of a Videhamukta. Vasiṣṭha says, "In men devoted entirely to knowledge. and even immersed in self-contemplation, arise that condition of "liberation in life," which is like the liberation following the dissolution of the body." He is a true Jīvanmukta, for whom the phenomenal world, wherein

3. J.M.V., pp. 12-15
4. Br. Up. (4.4.7)

he moves, and acts, ceases to exist, leaving alone the all pervading eternal noumena. A true Jīvanmukta neither flushes with joy nor fades with pain. He subsists on whatever comes of its own accord. He remains content with whatever goes to him in shape of alms etc. In the hight of his concentration he can have no awareness of garland of flower, sandal and the like. Even if, at intervals, concentration breaks, such awarness come back to him,. his strong discrimination keeps him off from avoiding or courting any of them. This leads hims to that condition which is free from pleasure as well as pain.

"He is the true Jīvanmukta. who is awake though asleep, who knows no waking, and whose knowledge is entirely free from vāsanā." (5-92 L.Y.V.). He is awake because, all his senses beginning with the eyes exist in proper places and have not ceased to act.

He is asleep because his mind knows no transformations. Waking here means perceiving objects through senses, does not apply to him any longer. When one has full knowledge the conceit of the knowledge of Brahman and the desire for enjoyment of any kind are the taints of his mind known as vāsanā. Hence freedom from vāsanā is attained by him as his mind knows no transformations. He is free from attachment, hatred fear etc. He remains pure within like sky, which though filled with smoke, clouds, dust and the like remains unaffected by them and remains always pure [5]

He is a real Jīvanmukta, whom the world fears not and who is not afraid of the world."

The world finds no cause to fear him for he never indulges in insulting others, nor the people insult him. He is at peace with the ways of the world. He is deeply immersed in all things yet keeps his head cool. Such coolness is attributed not only to his freedom from the worry of action, but also to his finding his complete self.[6]

5. J.MV.p.17
6. J.M.V.p.18

He is Sthitaprajña in Bhagvadgītā. The mind of one who has reached parāvairagya (Supreme detachment), through the practice of yoga, who has gained complete mastery over it. who has realised the Truth, can not be disengaged even for a moment from the Truth. This is a stable knowledge. He is sthitaprajna.

Again a sthitaprajña is he, whose mind is free from anxiety amid pains, is indifferent amid pleasures and is detached from passion, fear and anger, is called a sage of stable knowledge (B.G. II - 56), He, who on everyside, is without attachments, whatever hap of fair and foul, who has neither likes nor dislikes, of such a one the knowledge is well-poised (B.G.II - 57). When, again, as a tortoise draws in all sides its limbs, he widraws his senses from their objects, then is his knowledge is well-poised. (B.G.II - 58).

Thus mind of a sthitaprajña out of his concentration is entirely free from grosser kinds of transformation. On the other hand it can be said that when he is in a state of concentration, his mind is subject to no transformations. So there is no room for doubt as to their gross nature. So practice of concentration is necessary and helpful to control the mind, as well as at the realization of Brahman. As it is uttered by Lord Kṛṣṇa in Gītā The excited senses of even a wise man, though he be striving impetuously, carry away his mind, having restrained them all, he should sit harmonized, with 'Me' as his supreme goal, for whose senses are mastered of, his wisdom is well-poised. (B.G II - 60-61)

The true devotee of the Lord is described in the twelfth chapter of Bhagavadgītā.

'He who bears no ill will to any being, is friendly and compassionate without attachment and egosim, balanced in pleasure and pain , forgiving, evercontent harmonius with the self controlled, resolute, with mind and reason dedicated to Me, he my devotee is dear to Me'.

A person whom the world does not shrink away, who does

not shrink away from the world, who is freed from anxieties of joy, anger and fear, he is dear to Lord. He who wants nothing, is pure, expert, passionless, untroubled, renouncing every undertaking is dear to God. A liberated neither loves nor hates nor grieves nor desires, renouncing good and evil, who is full of devotion is dear to God. Thus a liberated surrenders his mind to God alone. He has balance of mind in all the time at any circumstances.(B.G.15 TO 19) such a balanced person i.e a liberated goes beyond three gunas i.e Sattva, rajas and tamas So he is a guṇantīta Service to Supreme by the practice of knowledge and contemplation; accompained by stable devotion are the means to be adopted by guṇātīta. Thus in Bhagvadgītā a liberated is described as sthitprajaña and a gunāntita.

In the Mahābhāṛata, Maharṣi Vyāsa describeed a liberated as Brāhamaṇa. The word Brahman refers to the knower of Brahman. He is a Paramahaṁsa.

"Dressed in patched garments and lioncloth holding a stick, intent on contemplation, who is alone in perpetual delight, Gods know him to be Brāhman".(Anuśāsana parva 251) Vidyāraṇya explains the śruti passage thus: The lion-cloth, the stick and the blanket, he should keep, as much for self-presentation as for obliging the world.' He should, however, never, even out of the strongest - sympathy, exchange even a word about the worldly concern (of his pupils), but should keep himself ever absorbed in contemplation The śruti has it thus—'know well that Ātman alone, avoid all other speech (Mun. Up. 2-2.5). The contemplation should meet with no interruption, in solitude

The mendicant should remain alone,two of them make a pair, three a village and more than that a city. A city, a village or a pair should be avoided. For there is scope for the discussion of political or other matters or gossip about alms. (Nārad up - 3-56-7).

'Him the Gods know to be a Brāhmaṇa who has no blessing

to give, no business to undertake, no salutation to offer nor praise to bestow, who is never depressed and in whose case rituals are shorn (of their purpose)' (Anuśāsana parva - 251).

'Gods know him to be Brāhmaṇa. who, all alone, ever fills the entire sky (ākāśa), and who finds himself in utter solitude, though in a densely populated place.' (Anuśasana parva 251).

Vidyāraṇya. then sets forth the characteristics of an ativaṇāśrami described in fifth chapter of Sūtasaṁhitā:

> 'An ativarnāśrami is beyond varṇa and āśrama who knows his self as God, from hearing the great texts of the Vedānta.'

'The Great Lord who is one, ever pervades, like sky (ākāśa) the body of the Cāndāla or of a Brāhamaṇa. of bird or beast and all things high or low, devoid of any connection whatever. The body, the senses, the vital breath, the mind, the intellect, egoism, thė thinking principle even illusion itself, the cosmos beginning with supreme ether, the doer, the sufferer, none of these constitute the self. It is all pure being, pure thought, all pure bliss. It is verily in every respect, none other than Brāhmaṇa .

Where all men are usually fast asleep, the yogin is wide awake, where men are wide awake there the enlightened one is fast asleeep. He is said to be a true ativarṅāśṛmi the one who is beyond Varṇa (cast) and āsrama (stage) (14 to 42).

Thus Śrutis call it as 'altogether librated is all the more liberated. The texts from the smṛitis speak of the liberated in life and sthitaprajaña, a bhakta, a guṇāntita, a brāhaman and also an ativarṇāśrami- all these bear testimony to the truth and possibility of what is called 'liberation in life' or 'Jīvanmukti'.

Vidyāraṇya also discusses the charactaristies of Videhamukti :- He says — As the air sometimes resumes its tranquility, stops its assumed motion, so the liberated self remains in his real essence leaving off his contingent empirical existence. The one who is liberated after death is beyond joy and sorrow. He is unaffected by either. He who is free from all conditions having his subtle body dissolved even here, can not be described as sat i.e. can not be called prajñā conditioned by Avidyā nor Īśvara conditioned by māyā — the root cause of the world experience.

In the view of the resemblance which Jīvanmukti is said to bear with Videhamukti, as long as there is the predominance of undifferenciated ecstasy in the Jīvanmukta, he is exalted.

Vidyāraṇya follows mainstream Advaita in holding that the fruits of currently manifesting karma obstruct ignorance destroying knowledge and cause the body and senses to continue. However he parts company with the Saṅkara's tradition by claiming that one can overcome the necessity of experiencing the fruits of actions by making personal efforts specifically yogic practices. Vidyāraṇya here arranges (unlike Śaṅkara) that even though prārabdha Karma is stronger than ignorance destroying knowledge, yogic practice is even stronger than Prābadha Karma. He repeatedly declared the necessity of human efforts to overcome being controlled by mental modifications and pre-existing Karma.

These efforts include Advaitic means such as following the sacred texts(śhāstras) and associating the wise as well as yogic efforts such as resisting desires and controlling mind. He approves the Laghu Yoga. Vāsiṣṭha claim that all results- such as hell, heaven or liberation- drive from human efforts, and only right effort can bring the highest goal. Śaṅkara. on the other hand, carefully separates the effort of yogic practices from knowledge which alone brings liberation.

Chapter - VIII

MEANS TO JĪVANMUKTI

PATH OF LIVING LIBERATION OR THE MEANS OF THE ATTAINMENT OF LIVING LIBERATION

Three-fold means :

[1] Obliteration of desires (वासना क्षय)

[2] Dissolution of mind (मनोनाश)

[3] Knowledge (तत्त्वज्ञान) is the subordinate mean; are the direct means to attain Jīvanmukti.

||Vāsanākṣaya||

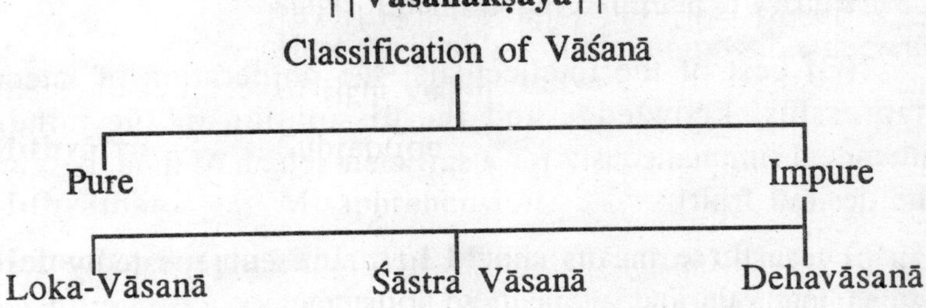

* Vāsanā in relation to mind
* Difference between Viṣaya-Vāsanā and manas-Vāsanā
* Process and means of obliteration of latent desires:

[a] By creating the qualities like friendliness and the rest

- Pleasure vs friendliness
- Hate vs compassion
- Virtue vs complacency & vice vs indifference

[b] By a discrimination

- Arrogance Begotten of learning (विद्यामद) and discrimination
- Anger and discrimination
- Vāsanā for wealth & discrimination
- Vāsanā pertaining to wife-children & discrimination

MEANS OF THE ATTAINMENT OF JĪVANMUKTI (वासनाक्षय)

The largest portion of the Jīvanmuktiviveka is devoted in describing and analysing the three-fold means for obtaining liberation while living. The three means according to Vidyāraṇya are Vāsanākṣaya. Manonāsha and Vijñana i.e. destroying mental impressions, extinguishing mind and knowing truth. Giving the reference of Laghu Yoga Vāsiṣṭa (28.1.116). Vidyāraṇya quotes:

वासनाक्षयाविज्ञानमनोनाशा महामते।
समकालं चिराभ्यस्ता भवन्ति फलदायिन:॥११६॥

(Of best of the intellectuals! the obliteration of latent impressins, knowledge and the dissolution of the mind, attempted simultaneously for a sufficient length of time, bestow the desired fruit).

These three means should be practised repeatedly and not at intervals and again these three factors are interlinked and interdependent on each other and can not be practised in isolation or one after another. Though these are the three means for attaining living liberation, Vāsanākṣaya and manonaśa are the principle means and the Vijñāna or knowledge is subordinate or the cause of the rest two means so it is an indirect means to living liberation.

This is the theme of Vāsanāksaya-Prakaraṇa and is the key to Jīvanmukti'.

Though Vidyāraṇya explains in detail how the repeated practice and interdependence of these means lead to Jīvanmukti. in the beginning of the chapter he describes that a mumukshu fails to attain his goal if he fails to practice these means continuously or repeatedly and simultaneously. Giving the references from Laghu Yoga.Vāsiṣṭa. (115 - 117) he describes the point by giving examples of the daily life He says

यथा संध्यावन्दन मार्जने सहविनियुक्तानाम् 'आपो हिष्ठा' इत्यादिना तिसृणामृचांमध्ये प्रतिदिनमेकैकस्या ऋचः पठि शास्त्रीयानुष्ठानं न सिध्यति यथा वा षडङ्गमन्त्राणाम् एकैक मन्त्रेण न सिद्धिः, यथा वा लोके शाक सूपौदनादीना-मेकैकेन न भोजनसिद्धिः, तव्दत् चिराभासस्य प्रयोजनमाह।

(i.e. if in the 'besprinkling' during the twilight' adoration ceremony; which must be accompanied with the repetition of the three Ṛgveda incantations, beginning with 'āpo hi ṣthā' etc, these three be repeated alternately day after day, the ceremony can not be said to be accomplished as laid down by the scripture. Again if from those incantations, which always go in hexads, one incantation after another were recited, at intervals, no result will follow, or, for instance, if, in our temporal affairs, vegetable, soup, rice and the like were each served, one for each meal. there can certainly be no dinner). He adds. 'Not only does isolation of these means in their application lead to no result, but also renders even their proper attainment impossible". By coupling together any two out of these three, there arise three pairs, in, each pair both the means are cause and effect of each other. Such as :

यावद् विलीनं न मनो न तावव्दासनाक्षयः।
न क्षीणा वासना यावद् तावाच्चित्तं न शाम्यति।।११०।।
यावन्न तत्त्वविज्ञानं तावच्चित्तशमः कुतः।

यावन्न चित्तोपशमो न तावत् तत्त्ववेदनम्।।१११।।
यावन्न वासनानाशस्तावत् तत्त्वागमः कुतः।
यावन्न तत्त्व संप्राप्तिः न तावद् वासनाक्षयः।।११२।।

Showing contrariwise relationship Vidyāraṇya quotes:

तत्त्वज्ञान मनोनाशो वासनाक्षय एव च।
मिथः कारणतां गत्वा दुःसाधानि स्थितानि हि।।११३।।

Vidyāraṇya analysis the relationship of these three means by discussing their positive and negative concomitance or the anvaya-vyatireka. Thus, He defines dissolution of mind. The inner sense (aṅtahkaraṇa) taking the form of a series of transformations, resembling, in this respect, the flame of a lamp, is called the mind on account of its mental functions. The dissolution of this mind then means its giving up all connections, with these transformations and assuming a state of interception. Giving the reference from Yogasūtra' of Patañjali he quotes: "Interception is the transformation of the mind. At the moments of interception; the impressions of distraction and interception going out and rising up respectively (111-9). impressions tending to distraction being neutralized, impressions holding the mind in control manifest themselves, and the moment of interception holds sway over the mind. This is all that is meant by the **"dissolution of the mind"**. That which generates mental impressions, such as anger and the like, rising up all of a sudden and without any thought of the past or future, is called vasanā (or latent impressions), for it is something that saturates the mind with the vāsanā (lit. - smell or fragrance) of every previous exoerience. The obliteration of these latent impressions means the absence of scope for the rise of anger and the like, even in the presence of appropriate external causes, as a result of pure vāsanās, such as self control and the like, generated by the discrimination being firmly fixed in the mind. Now if the mind is not dissolved and transformations do arise, there can be no obliteration of vāsanā. in case some accidental external cause excites anger and the like. On the other hand, as long as vāsanā is not destroyed, transformation of the mind continues as before and as such there is no 'dissolution of the mind'.

Same is the case with knowledge and latent impressions. Knowledge never comes about, for want of proper means, such as self-control and the like, as long as the latent power of anger and the like is not laid entirely at rest. In so far as the secondlessness of Brahman is not fully realised, that false knowledge, which invests the causes of anger and its brood with a semblance of reality, is not done away with: hence latent impression is not killed (J.M.V., p. 38-39).

Vidyāraṇya then proceeds to set forth the reciprocal causal relationship of the aforesaid three pairs in their positive aspect. He says: "When the mind is dissolved and there is no conscious-ness of any external cause which can fully rouse mental impression, latent impressions fade away. When latent impressions dies out and there remains no cause for that functioning of the mind, which we call anger and the alike, the mind is dissolved. In the same way when knowledge has arisen, as the mind ceases to think of this unreal world of the 'horns of a man' and similar impossibilities, and as thinking in any form is no more of any use, after the Ātman has been fully seen the mind dies out, of itself, like fire not fed with fuel. This was about the pairs of dissolution of mind and obliteration of latent desires' as well as knowledge and obliteration of latent desires and knowledge and the dissolution of the mind.

रिपौ बन्धौ स्वदेहे च समैकात्म्यं प्रपश्यत।
विवेकिनः कुतः कोपः स्वदेहावयवेष्विव।।२.१८।।
गुणाः शमादयो ज्ञानाच्छमादिभ्यस्तथा ज्ञाता।
परस्परं विवर्धेते व्दे पद्मसरसी इव।।४.१०७।।

(J.M.V., p. 40, Naiṣkarmyasiddhi)

(Wherefore can the man of discrimination find cause for anger, when he sees the identical Ātman pervading his foes, kinsmen and himself, alike, even as it does the limbs of his body ? 2.18 Nai)

(Virtues such as self-control and the like, derive nourishment from knowledge ; knowledge likewise developes fully through these virtues like the lotus and the pond wherein it grows, they conduce to their mutual well being. (4.107)

In the Manusmriti, it is rightly said that:

[i] 'Personal efforts' connotes requisite enthusiasm to form the firm resolve 'somehow or the other I shall varily accomplish this.

[ii] Discrimination' means the conviction borne on one, after due analysis and observation, that śravan, manana and nididhyāsana are the means to knowledge.

[iii] Yoga is the means to the dissolution of the mind, and the setting up of an opposite current (of vāsanā) is the means to the obliteration of vāsanā. It is said that the desire for enjoyment has to be kept "at a distance', because once the smallest desire for enjoyment is admitted, there will be nothing to prevent it from exceeding the limit, as borne out by the adage. This will continuously increase like fire constantly fed with fuel'. (Manu. II 94)

The nature of the subsequent application of the knower must be thus: It is the frequent rememberance of the sublime truth through whatever means. To quote the verses from L.Y.V.:

तच्चिंतनं तत्कथनमन्योन्यं तत्प्रबोधनम्।
एतदेकपरत्वं च ज्ञानाभ्यासं विदुर्बुधाः॥१०८॥
सर्गादावेव नोत्पन्नं दृश्यं नास्त्येव तत्सदा।
इदं जगदहं चेति बोधाभ्यासं विदुः परे॥१११॥
अत्यन्ता भाव संपत्तौ ज्ञातुर्ज्ञेयस्य वस्तुनः।
युक्त्या शास्त्रैर्यतन्ते ये ते तत्राभ्यासिनः स्थिताः॥११०॥

Thought of that, talk of that, mutual enlightenment about that and being intent ever on That - this the wise call the practice of knowledge.

The objective world was not in existence at the commencement of creation nor did it made up of 'I' and 'this world' ever exist - the wise know this as the best form of the practice of awareness.

They are verily the students who with the aid of the

scripture and skillfull practices, try to realise the absolute non-existence of the knower and the known (L.Y.V. -6., 108, 111, 110)

Mental manifestations cease by knowledge, i.e. knowing the apparent diversity is illusory and the self alone it real. Until non-duality is known, mental forms continue as a fire constantly fed with fuel. Knowing the truth of non-duality destroys mental impressions, removes bondage and reveals non-dual Brahman.

The knowledge of the falsity of the knower and the known is the knowledge of their non-existence and the non-experience even of their forms, is the realization of their absolute non-existence'. Skillful practices' refer to 'yoga'. This implies application to the dissolution of mind'.

In this way, when love and hatred are reduced to the utmost thinness, through the realization of the non-existence of the objective world, there palpably arises a sense of pleasure This condition is called the study of Brahman (6.112).

Bondage according to Vidyāraṇya :

Vidyāranya believes that all the three means appear to be equally important, but for some one is principle and the other is subordinate for some other vice-versa To one desirous, the two objects are 'Jīvanmukti' and 'Videhamukti. Sruti says: (kath, Up. 2.1.1):–

विमुक्तश्च विमुच्यते। इति श्रूयते तत्र जीवतः पुरुषस्य दैवसंपदा मोक्षः, आसुरसंपदा बन्धः।

the already liberated is altogether liberated. Though life of the higher self, the living man attains liberation Through life of the lower self bondage.

This has been proved by giving six verses from Bhagavad-Gītā

दैवी संपद्विमोक्षाय निबन्धाया सुरी मता।।१६.५।।

The divine properties are deemed to be for liberation, the demonical for bondage (16, 5).

These two lives are described in this way; अभय (fearlessness), (cleanness of life),सत्वसंशुद्धि, ज्ञान-योग व्यवस्थिती (steadfastness in the yoga of wisdom), दान (almsgiving), दम (self-restraint),यज्ञ (sacrifice) and स्वाध्याय (study of scriptures) तप (austerity: and शान्तिदया भूतेष्वलोलुप्त्वं (straightforwardness) आर्जव, अहिंसा, सत्य, क्रोधत्याग compassion for living beings) मार्दवं (uncovetousness) तेजः क्षमा, धृतिः etc. are the qualities, one who has divine properties; whereas Hyprocrisy (दम्भ) (arrogance) दर्प अभिमान (conceit) क्रोध (wrath) and पारुष्यम् (harshness) अज्ञान (unwisdom) are the qualities who is born with demonical properties (J. M.V. p. 43).

The point is that when the evil vāsanās born of the 'Life of the lower self, inherent in one from birth and running counter to the teachings of sacred texts, we put out by the good vāsanās brought about from the life of a higher self accomplished by personal efforts and carried on in accord with the word of scriptures. There results 'Jīvanmukti'.

The dissolution of the mind is also mentioned in the Śruti as the cause of Jīvanmukti :

मन एव मनुष्याणां कारणं बन्धमोक्षयोः।
बन्धाय विषयासक्तं मुक्त्यै निर्विषयं स्मृतम्।।२।।
यतो निर्विषयस्यास्य मनसो मुक्तिरिष्यते।
अतो निर्विषयं नित्यं मनः कार्य मुमुक्षुणा।।३।।

Mind alone is the cause of man's bondage or liberation, lost in enjoyment, it leads to bondage, freed from the objective sphere it leads to liberation As mind freed from the objective sphere leads to liberation, one desirous of liberation must always try to wipe off the objective sphere from the plane of his mind

Vidyāraṇya further says that this bondage is of two kinds :

Tivra and Mrudu, i.e. fast and loose. The life of the lower self being the direct cause of misery makes up the former kind of bondage. The mere sense of duality, not being by itself, the cause of misery and for the reason that it leads only to the life of the lower self, is the other kind of bondage. Whereas only the former kind of bondage is removed by the obliteration of latent impressions, both kinds are removed by the dissolution of the mind.

Thus, Vāsanākṣaya and Manonāśa are the direct means of living liberation. Both are the principle means where as knowledge is subordinate, being only a mediate cause as producing the first two. The Śruti mentions knowledge as the cause of Vāsanākṣaya in various places:

'ज्ञात्वा देवं सर्वपाशापहानि:।' (Śve. 3.1.1.11)

i.e knowledge of the Divine dissolves all bonds

'अध्यात्म योगाधिगमेन देवं मत्वा धीरो हर्ष शोकौ जहाति।'

(Kaṭha. 1.1.12)

(The wise one realizing the effulgent one through the yoga of Ātman, gains freedom from joy and sorrow) (Katha 1 1-12).

'तरति शोकमात्मवित्'

(i e. The knower of the Ātman transcends all miseries)

'तत्र को मोह: क: शोक एकत्वमनुपश्यत:।'

(What is delusion, what sorrow when one realises oneness?

(Īśa. 7)

'ज्ञात्वा देवं मुच्यते सर्वपाशै:।' (Śve Up. 1.1.8)

(He is freed from all bonds whatever, on knowing the effulgent)

Knowledge conduces to 'the dissolution of the mind' has also the sanction of the Śruti. Relating to the condition of knowledge is it said: 'In the case of one who sees only the self in all, what can he see with what, what can he smell with what?

(Br. 2.4.14)

As Shri Gaudapādācārya says :

आत्मतत्त्वानुबोधेन न संकल्पयते यदा।
अमनस्तां तदा याति ग्राह्याभावे तदग्रह:।।३.३२।।

i.e. when the mind does not bring forth any more of these imaginations because of the knowledge of Truth, which is Ātman (pure consciousness), then it ceases to be. Mind becomes free from the idea of cognition for want of objects of cognitions. Thus, Vāśanākṣaya and manonāśa are the principal causes of Jīvanmukti whereas knowledge is the principal cause of Videhamukti (liberation after death), it is the direct means of attaining the same. As Smriti says:

ज्ञानादेव तु कैवल्यं प्राप्यते येन मुच्यते। इति स्मृतेः

From knowledge alone is attained 'Kaivalya' where with one is liberated.

Kaivalya is the aloneness of the self, to wit, being without the appendage of the body and the like This is attainable by knowledge alone; for, the condition of corporeality, which is the outcome of ignorance, is capable of being dissolved only through, knowledge. The word 'alone' added to qualify knowledge is meant to exclude formal activities (Karmas) As Sruti says: न कर्मणा न प्रजया न धनेन (Kai.Up. 4. 12.3) neither by Karma nor by Progeny nor by wealth'. He who without properly studying the philosophy of knowledge brings about the dissolution ot the mind and the obliteration of latent impressions as occasion arises therefore and devotes himself to Brahman with characterstics never realises Kaivalya for his subtle body is not destroyed. Hence the above two (by themselves) are also excluded by the word 'alone'. The words 'wherewith one is liberated' mean that by Kaivalya brought about by knowledge, the individual is freed from all bonds and conditions whatever. Bonds are of various kinds, signified by the expressions, the knots of Avidyā, the conviction 'I am not Brahman', the knot of the heart, doubt, Karmans, hankering after all objects of desire, death, rebirth, these and others wherever they occur. All the aforesaid bonds can be removed by knowledge. Valuable quotations from various Srutis proves this:

(1) एतद्यो वेद निहितं गुहायां सोऽ विद्याग्रन्थिं विकिरतीह सोम्य।

(Muṇḍak Up. 2.2.10)

(O good one! he cuts asunder the knot of Avidyā, who finds it ever present in the cavity (of the heart))

(2) ब्रह्मविद् ब्रह्मैव भवति। (Muṇḍak Up. 3.2.9)

(He becomes Brahman who knows Brahman)

(3) भिद्यते हृदयग्रन्थि छिद्यन्ते सर्वसंशयाः।
क्षीयन्ते चास्य कर्माणि तस्मिन् दृष्टे परावरे।। (2.2.8)

(The knot in the heart is cut asunder, all doubts vanish, all karman fades away, on the realzation of the Supreme. 2.2.8)

(4) यस्तु विज्ञानवान् भवति समनस्कः सदा शुचिः।
स तु तत्पदमाप्नोति यस्माद् भूयो न जायते।। (Kaṭha 1.3.8)

(He who has acquired knowledge, who is in the right frame of mind and ever pure finds that status whence he is not born again.

(5) य एवं वेदाहं ब्रह्मास्मीति स इदं सर्वं भवति। (Br. Up. 1.4.10)

(He who thus realizes "I am Brahman", becomes this all) (Br. Up. 1.4.10)

By Videhamukti, Vidyāraṇya means release from future embodiment. Thus, knowledge is the principal cause of Videhamukti. Several people also maintain that videhamukti comes after the dissolution of the existing body. To quote from Śrutis.

"He tarries only so long as he is not liberated; the moment he is liberated, he becomes one with the All" (Chān. Up. 6.14.2)

Even the author of Vedāntasutras says: "He becomes It, having spent the other two through experience" (IV 1.19). Meaning by 'the other two' the previous good or evil karmas'.

So also Vasiṣṭha says :

जीवन्मुक्तं पदं त्यक्त्वा स्वदेहे कालसात्कृते।
विशत्यदेहमुक्तत्वं पवनोऽ स्पन्दतामिव।। (L.Y.V. 5.98)

("After leaving the condition of the 'Jīvanmukta'. one enters on the condition of the Liberation after death on the disintegration of ones body by lapse of tenure, even as the wind gets becalmed LY.V. 5.98)

Though the above two positions appear to be different both opinions are possible depending as they do on the point of view of the speaker and either does not necessarily contradict the other The word 'deha' contained in the term "videhamukti has been understood as referring to all kinds of bodies by many, in the description given by them there of Vidyāraṇya uses the word only in the limited sense of future body i.e. embodiment after the disappearace of the present body).

Acquisition of knowledge prevents future emobdiment. Knowledge expels ignorance only. The means of dissolving the subtle body is the cessation of operating causes. An effect lapses in one or two ways: By the presence of its opposites, or by the cessation of the operating cause, even as the lamp goes out, either when its adversary, namely oil and the wick. We do not see any direct opposit of the subtle body. Operating causes are of two kinds: Kinetic and potential. It is because of these two that the subtle body of the ignorant exists in this world and the next. But in the case of Enlightened, potential causes are removed by knowledge and Kinetic causes are removed by (their being spent up in) actual life; and their subtle body goes out like the lamp deprived of oil and the wick. Hence it is plain that such dissolution (of the subtle body) is not the result of knowledge (J.M.V. p. 47, 48).

Arguing the same strain, Vidyāraṇya further says that the prevention of the further embodiment also is not the result of knowledge. Here the question may arise "Is it the non-existence prior to embodiment or the prevention of such non-existence that is the result of knowledge?' The first alternative can not hold good; for such antecedent non-existence has no beginning. Nor can the second alternative hold water. The prevention of the antecedent non-existence of the future embodiment can be brought about simply by the cessation of the operation of the potential karma Nor can the total doing away with future embodiment be the result of knowledge for knowledge destroys ignorance alone.

Thus. '*Videhamukti* in the sense of preclusion of future embodiment, is simultaneous with the rise of knowledge in Br. Up. *Yājñāvalkya*'s words are to the same effect:

'अभयं वै जनक प्राप्तोऽसि।' (Br. Up. 4.2.4)

(O Janaka you have reached the condition of *supreme fearlessness*)

एतावदरे खलु अमृतत्वम्।' (Br Up. 4.5.15)

This verily is *real immortality*.

'तमेवं विव्दानमृत इह भवति।' (Taitt. Ār. 3.12.7)

(Who knows Him thus, becomes immortal even here)

If Videhamukti. the result of knowledge should not immediately follow the rise of knowledge, but should come about after sometime, it will be necessary to imagine some other thing known as 'Apurva' derived from knowledge (as intermediate between the two) even as it is necessary to do so in the case of Jyotistoma and other sacrifices. So the whole of the philosophy of knowledge (Jñāna-śāstra) would be included within the province of ritualistic lore (Karma-śāstra). At the end of this point, Vidyāraṇya concludes that:

Let the 'Videhamukti' (of the form of liberation after death), which you hold dear, come after the dissolution of the present body: but the videhamukti (Liberation from future embodiment) we cherish is surely simultaneous with the rise of knowledge. The Lord Śesa (शेष) too says with this idea in mind:

"Whether in some holy place or in the house of a cāṇḍāla, though bereft of consciousness, he stands liberated in the very moment of the rise of knowledge and finds Kaivalya on giving up the body, with the cause of sorrow - entirely removed (Paramārthsāra 81) (J.M.V., p. 49)

Thus is established the premier position of knowledge which is the direct means of attaining videha-mukti. while 'the obliteration of latent impressions and 'the dissolution of mind occupy only a subordinate position therein. It being merely the means of acquiring knowledge. That the life of the higher self, which brings about the destruction of self, which brings about the destruction of the impressions of the life of the lower self, is the means of acquiring knowledge is borne testimony to by Śrutis and Smṛitis alike.

In the Śruti it is quoted as:

'शान्तो दान्त उपरतस्तितिक्षुः समाहितो भूत्वात्मन्येवात्मानं पश्येत्।

(Br. Up. 4.4.23) J.M.V..p.50

(After becoming self restrained, self controlled indifferent forbearing and one pointed one should find the self in onself).

The smṛiti also says :

(Humility, unpretentiousness, harmlessness, forgiveness rectitude service of the teachers, purity, steadfastness, self control, dispassion towards the objects of the senses, and also absence of egoism, innsight into the pain and evil of birth, death, old age and sickness, unattachment, absence of self-indentification with son, wife or home, and constant balance of mind in wished for and unwished for events, inflinching devotion to Me by yoga without other objects resort to sequestered places, absence of enjoyment in the company of men, constancy in the wisdom of the self understanding of the object of essential wisdom - all this is declared to be the wisdom, what is other than this is ignorance. (B.G. 13. 8 -12).

Thus is established the principle and subordinate relation, one to the other of the triad Vāsanākṣaya and manonāśa from the stand point of Jīvanmukti and Videhamukti.

Classification of Vāsanā:

Vidvāraṇya states that Vāśanā is of two kinds. They are

pure and impure. To prove this he quotes some verses from Yoga Vasiṣtha where sage Vālmiki says:

वासनाद्विविधा प्रोक्ता शुद्धा च मलिना तथा।
मलिना जन्म हेतुः स्याच्छुद्धा जन्मविनाशिनी॥१०॥
अज्ञान सुघनाकारा घनाहंकारशालिनी।
पुनर्जन्म करी प्रोक्ता मलिना वासना बुधैः॥११॥
पुनर्जन्मांङ्कुरं त्यक्त्वा स्थिता संभृष्ट बीजवत्।
देहार्थ ध्रियते ज्ञातज्ञेया शुद्धेति चोच्यते॥१२॥ (L.Y.V. I. 10, 11, 12)

(Vāsanā is of two kinds pure and impure, the later leads to frequent (L.Y.V. (10, 11, 12) incarnation, the former puts and end to all incarnation whatever. The second kind of vāsanā is described by the wise as of the form of extra dense ignorance, made up of intense egoism, bringing about birth and death in secession. The other, like seeds sown after being baked on fire, does not send out the sprout of reincarnation but continues only for the sake of the body, having known the thing to be known).

Vāsana arises due to ignorance. Ignorance is that which veils the distinction between the five sheath comprising the physical body (annamaya) and others and the transcendent witness of them all i.e. cid Ātman. Vidyāraṇya explains this point neatly. He says: As milk is curdled on being mixed with butter milk, or as melted clarified butter is congealed on being exposed for a long time in a cool place, so does this vāsanā (through ignorance). Denseness in this instance refers to the series of delusion brought about by vāsanā. In the sixteenth chapter of Bhagavad Gītā, Lord refers to this vasana, while explaining the life of a lower self.

Demonical men know neither right energy (प्रवृत्ति) nor right abstinence nor purity nor even propriety, nor is truth in them. "The universe is without truth, without basis". They say "without God'. Holding this view, these ruined selves of small understanding of fierce deeds come forth as enemies for the destruction of the world. Surrendering themselves to insatiable desires, possessed with vanity, conceit and arrogance holding evil ideas through delusions, they engage in actions with impure resolves. Giving themselves over to unmeasured thought whose end is death, regarding gratification of desires as the highest, feeling sure that this is all held in bondage by a hundred ties of expectation, given over to lust and anger, they strive to obtain by unlawful means hoards of wealth, sensual enjoyment. (B. Gītā 16.7-12)

Pure Vāsanā, however, is of that kind which "knows the thing to be known'. The form of the thing to be known is thus set forth by the Lord in chapter 13 of the Bhagvad Gitā.

I will declare that which ought to be known that which being known, immortality is enjoyed The beginingless. supreme, eternal, called neither being nor non-being Everywhere 'THAT' has hands and feet, everywhere eyes, heads and mouths all hearing THAT dwells in the world enveloping all. Shining with all dwells sense faculties, without any senses; unattached, supporting everything and free from qualities, enjoying qualities without and within all beings, immovable and also movable by reason of IT'S subtlety, imperceptible at hand and far away is 'THAT' etc (B.G. 13. 12-17).

Thus are set forth conditioned and unconditioned forms (of Brahman) with a view to facilitating the comprehension of its tatastha laksaṇa and swarūpa laksaṇa.

Vidyāraṇya still makes it clear that pure vāsanā is that activity of the senses which is still pervaded with an undercurrent of the perception of the thing: It is of use only in keeping the body alive. It leads neither to the life of the lower self consisting of hypocrisy, vanity, and the rest, nor to the production of dharma and adharma (religious merit or demerit) leading to future incarnations, Grains of corn such as paddy and the like, baked on fire, may be stored in a granary, however; they can of no use either for delicious food or for producing any fresh corn. (J.M.V., p. 56)

Three fold division of Impure Vasanā :

The impure vāsanā is further classified into three : [1] Loka-vāsanā i.e. vāsanā relatng to the world. [2] Śāstra-vāsanā i.e. vāsanā regarding one's own learning when causes vanity and pride in man [3] Deha-vāsanā. All these impure vāsanās obstruct the path of knowledge in its established form. Both Śrutis and Smṛtis suggest that the enlightened should give up all such vāsanās. These three vāsanās then relating to the world, learning, and the body, though apparenty agreeable to the mind of the uniformed should entirely be given up by those who have acquired discrimination: for they obstruct the rise of knowledge in the seekers and interfere with the permanance of knowledge acquired by the knower (J.M.V., p. 62).

Vāsanā in relation to mind :

The mind or Chitta is the form of three qualities of prakṛti viz. Sattva. Rajas and Tamas as radiance (Prakāsa) outgoing energy and delusion (Moha) are to be seen there. In Bhagavadgītā it is said:

'सत्वात् संजायते ज्ञानं रजसो लोभ एव च।
प्रमादमोहौ तमसो भवतोऽज्ञानमेव च॥'

(From sattva wisdom is born and also greed from rajas; heedlessness and delusion are of tamas and also unwisdom) (B.G. 14.17)

But the guṇas constantly change their form one after another in a series, like the waves of the ocean, and any one of them gain mastery over the rest. When tamas predominates, there arises the life of the lower self; the predominance of rajas gives sway to the three vāsanās, relating to the world etc; when sattva gains mastery, the life of a higher self is established (J.M.V., pp. 64-65)

In Bhagavad Gitā>, it is described in the following words:

सर्वव्दारेषु देहेऽस्मिन् प्रकाश उपजायते।
ज्ञानं यदा तदा विद्याद् विवृद्धं सत्वमित्युत॥ (14.11)

(When the wisdom light streams forth from all the gates of the body; then it may be known that Sattva is increasing)

Though the mind appears to be endowed with all three properties, its principle material cause is sattva alone. Rajas and tamas only are accessories. Hence it is that Sattva is the residual native form of the mind of a man of knowledge, when he has done away with rajas and tamas, through application of yoga. Hence this sattva is fit to receive the height of Ātman. Śruti says: It is seen with the one pointed and subtle intellect by men of subtle wisdom (Kaṭh. 1. 3.12). Thus is set forth the nature of vāsanā" and manas.

The way to get rid of viṣaya-vāsanā and mano-vāsanā:

The Vāsanā arising from the predominance of Tamas, is called Viṣaya-vāsanā, relating to the object of enjoyment, which refer to the life of lower self consisting of hypocrisy, pride and the alike. This is the most impure vāsanā. The other impure vāsanā is called manas-vāsanā.

Now both these vāśanās should be eradicated alike. Mind should be devoid of all these. The mind without them is pure sattva. Vidyāraṇya firmly says that though vāsanās are not something like dust and straw which can be swept with brush, we can take vow to give up them. The vow here consists in uttering the formula of renunciation (praiṣa) after which .you may keep the vow with all possible care. Those who can not pronounce the Vedic formula may take the vow in .their own vernacu'ar. If you say that in fasts we do not touch vegetable, soup, rice and the like, here too there is avoiding of all contacts with flowers, perfumes and the like (J.M.V., p. 67). Whereas you have in the first instance, pastimes provided in-the form of listening to the exposition of the purāṇas. worshiping gods, dancing, singing, playing on musical instruments and the like. All these tend to draw the mind away from hunger, sleep or sloth.

Vāsanās can be kept away by creating friendship and the rest in the mind :

This is mentioned by Patañjali in his aphorism:

'मैत्री करुणामुदितोपेक्षणां सुखदुःख पुण्यापुण्य विषयाणां भावनातश्चित्त प्रसादनम्।' (1.33)

(The mind becomes clarified by the practice of friendship, compassion, complacency, and indifference respectively towards happiness misery, virtue and vice) (1 33). The mind is muddled by love and hate, virtue and vice hankering after the pleasures muddles mind. If the hankering is put to an end, the mind becomes transparantly clear, like a river during the fall (Śarad) season (JMV, p. 68)

Anger and discrimination :

The practice of discrimination destroys anger the impurest of all other vāsanās. It is said to be twofold viz. Anger in one's self for others and anger in others pertaining to one's self. Both are the merciless consumers of their own seats, the person concerned. They serve no useful purpose and are in no way conducive to well bring either in this world or in the next. Hence the sādhaka should recognise their follies and faults and should overcome them. If once their hollowness is recognised, they stand no more."[1]

Vāsanā for wealth, pertaining to wife and children:

The best way to overcome the vāsana for wealth is to realize its nothingness and transitoriness. One should think that 'there is considerable trouble involved in the acquisition of wealth, likewise in its preservation. This wealth produces misery in everyway.' This sort of discrimination will make men aware of evils of running after it. Vāsanā characterising the attachment to wife and children also can be overcome if we come to discriminate that the vāsanā for their possession give nothing but pain.[2]

Thus we see that by cultivating good qualities and by discrimination all our impure vāsanās are overcome.

---

1. Ad. Concept J.M.: L.K.L. Shrivastava p. 198
2. Ibid p.199

Transformation of the mind also is necessary for its purification :

When the mind takes the form: "May all such pain be never mine'. This feeling of pain is, however, not easy to get rid off; as long as disease, enemies, tigers and the like continues to exist; nor is it entirely possible to do away with the cause of pain. Thus, then, the feeling of hatred or dislike consumes the heart like fire. When through the exercise of compassion for every miserable being, one feels that others like oneself should never experience pain of whatever kind, the mind becomes clear. The hatred felt towards the enemies and the like vanishes. It has been said: Life is as dear to all beings as it is to onself; the wise feel compassion for every being, taking their own selves as the standard (of comparison) (J. M. V. p. 68)

Great men show how this can be achieved:

A person having pure mind always thinks :

सर्वेऽत्र सुखिनः सन्तु सर्वे सन्तु निरामयाः।
सर्वे भद्राणि पश्यन्तु मा कश्चित् दुःखमाप्नुयात्।।

(May all be happy in the world, may all enjoy immunity from disease; may all find themselvs at their best let no one come to grief) Thus, the feeling of betterment of all beings helps one to get rid of vāsanās)

But all beings, by nature do not lean towards virtue, but are prone to vice. Men love the benefits of virtue but do not act virtuously: while abhorring the wages of sin. they persist in committing it.

These acts virtuous and vicious, both lead to remorse. If one feels exultant at the sight of virtuous men. one would, out of such vāsana (habit) turn, of one's own accord to act of virtue with a clear conscience Similarly, if one shows suprme indifference towards the sinful, one easily keeps oneself off sin. Thus, there being no cause for remorse, the mind gets clarified.

The feeling for friendliness for those who are happy, there follows disappearance of malice, jealousy and the like. When out of friendliness, the happiness of others, is made our own, how can there be malice and the like? By cultivating the habit of feeling, compassionate towards the miserable, the feeling of hatred, which culminates in the destruction of enemy and the like goes away from one (J.M.V., p. 69).

Vidyāraṇya also holds discrimination as the means to overcome impure vāsanās :

Since man is not capable of cultivating all pure vāsanās to counteract impure ones, he should by discrimination, examine his own mind and see how much impure vāsanās are there in mind He says, 'As one harassed by children, friends, wife and the like, feels much disgust for them all and he takes himself to renunciation, so should one, affected with the impure vasanās, such as arrogance begotten of learning, Opulence, pedigree and the like cultivate discrimination to counteract them.

(यथा पुत्र मित्र कलत्रादिभि: पीड्यमानस्ततो विरक्तस्तन्निवर्तकं पारिव्राज्य गृहाति तथा विद्यामद धनमद कुलाचारमदादि मलिनवासनभि: पीड्यमानस्तद्विरोधिनं विवेकमभ्यस्येत्। (J.M.V., p. 71

One may ask here. What is the use of cultivating discrimination, after the knowledge of Brahman is attained? It is said to be one of the four prerequisites of attaining Brahman knowledge (the four pre-requisites according to the teachers of Advaita-vedānta are [1] Viveka. [2] Vairagya; [3] Satsampatti; and (4] Mumukṣutva) To this Vidyāraṇya replies that the discrimination, which succeeds Realization is aimed at for acquiring peace of mind. He holds that knowledge and peace are two different phenomena. The knowledge of Brahman comes to one only after the mastery of the above four means. This is the reputed royal road common to all men. In the case of Janaka, however through the maturity of previous merits. Knowledge dawned upon him suddenly as a result of his having heard the 'Siddha-Gītā' even like a fall of a fruit from heaven. Thence discrimination had to be resorted to, for acquiring peace of mind.[3]

3. J.M.V., p. 72

Vidyāmada and discrimination :

Vīdyāraṇya says that there is possibility of occuring impure vāśanās like arrogance begotten of learning, anger and the rest, after the dawn of knowledge as is evidenced in the case of Yajñāvalkya and Bhagiratha: "Even the reputed knowers of Brahman like Yajñāvalkya and others were subject to the influence of impure vāsanā[4]. The arrogance begotten of learning is two fold. viz. Subjective and objective. The former as in the case of Śvetaketu is the pride felt by one as superior to others It will be overcome if one is constantly having the idea before his mind that his pride will be crushed by someone, as it was happened with Śvetaketu[5]

The objective pride is that. which others feel and which tend to make us feel our inferiority at every step. Now the opposite course to curbe this (pride) is to think. The other man is simply beside himself with pride; let him slander or insult me, either way I can not come to grief.

This is the cause that an enlightened saint is unmoved by slander and the rest.

(Having restrained them all, he should sit harmonised with Me as his supreme goal; for whose senses are mastered, of him the understanding is well poised.)

(Therefore, O mighty armed! Whose senses are all completely restrained from the objects of senses, of him the understanding is well poised) (Bhagavad Gītā 2. 61-68).

Vidyāraṇya is of the opinion that if a man cultivates the vāsana of pure knowledge in his mind, then it alone is sufficient to do away with impure vāsanā. It is possible only due to dhyāna and samādhi. He quotes so many verses from Laghu Yoga Vāsiṣṭa to prove that a Jīvanmukta must be free from all vāsanās because bondage is due to vasanās and void of vāsanā is liberation. As Vasiṣṭha says :

---

4. Ibid p.72
5. Ibid p.75

तस्माव्दासनया बद्ध मुक्तं निर्वासन मनः
राम निर्वासनीभावमाहराशु विवेकतः।।
सम्यगालोचनात् सत्याद् वासना प्रविलीयते।
वासनाविलये चेतः शाम्यत्यस्नेहदीपवत्।। (L.Y.V. 16.45.46)

यो जागर्ति सुषुप्तिस्थो यस्य जाग्रन्न विद्यते।
यस्य निर्वासनो बोधः स जीवन्मुक्त उच्यते।। (L.Y.V. 5. 62)

(The mind, therefore, experiences bondage through vāsanā; void of vāsanā is liberation O Rama! Try as soon as possible, to acquire freedom from vāsanā. Vāsanā melts away, on developing the right type of intution. On the dissolution like the flame of an oil-less lamp.

He is the real Jīvanmukta who is wide awake though asleep, who does not know any waking and whose knowledge is ever free from all vāsanās.

Thus, in the vāsanākṣaya prakaraṇa, Vidyāraṇya concludes that: It is plain that on the obliteration of desire, in the manner described, Jīvanmukti of the kind explained herein, immediately follows !

Chapter - IX

DISSOLUTION OF MIND

* Definition and nature of mind
* Means for the Dissolution of mind
* Two-fold yoga
* Eight-fold yoga
  - [a] Restless stage of Mind (क्षिप्तावस्था)
  - [b] Torpid stage (मूढावस्था)
  - [c] Distracted stage (विक्षिप्तावस्था)
  - [d] Concentrated stage (एकाग्रावस्था)
  - [e] Controlled stage or (निरोधावस्था)
* Process of attaining the unconscious variety of trance
* Obstacles in the way of Absolute Interception

Knowledge as the means to Jīvanmukti

* Knowledge as the cause of obliteration of latent desires
* Knowledge as the cause of Dissolution of the Mind
* Inter-relation of the three means

THE DISSOLUTION OF MIND (मनोनाश)

Dissolution of mind' is the second (principal) means to attain Jīvanmukti. In fact when the vāsanās are obliterated in one, one's mind is obviously dissolved with it, yet, one should carefully devote oneself to the dissolution of mind to the exclusion of everyother thing, further immunity from vāsanās once obliterated will be secured, thinks Vidyāraṇya. To support his view, he quotes some verses from Loghu Yoga Vāsiṣṭa where Janaka Says:

"The mind is, as it were, the root of the tree of saṁsāra (the wheel of birth and death), with its innumerable sprouts, branches, fruits and leaves. I believe it to be none other than imagination. By ceasing to imagine, I parch the mind up so as to parch up the tree of saṁsāra. I am awakened! I am wide awake! I have now found out the thief, who has stolen my self. It's name is mind. I shall presently inflict the extreme penalty on him; I have been subjected to severe torcher all along by him (LYV. 19. 53.5)[1]

Thus, mind being the root cause of all attractions, it is necessary to control one's own mind. As sage Vasiṣṭha has said "The rising of the mind brings on misery in its wake, while the dissolution of it brings on happiness; the mind of a wise man dissolves in no time, while that of the ignorant forms, as it were, his fetters (LYV 17.5)

On the face of this wide earth, it is only those enlightened souls, who are not subduced by their own mind, that are really happy and deserve to be numbered among heros (LYV 15-19).

In the case of Yogins in genera!, the condition of fearless-

1. सहस्त्राङ्कुर शाखात्मफलपल्लव शालिनः।
अस्य संसार वृक्षस्य मनोमूलमिति स्थितम्।।
संकल्पमेव तन्मन्ये संकल्पोपशमे न तत्।
शोषयामि यथा शोषमेति संसारपादपः।।
प्रबुद्धोऽस्मि प्रबुद्धोऽस्मि दृष्टश्चोरो मयात्मनः।
मनोनामेह हन्म्येनं मनसास्मि चिरं हतः।।
मनसोऽभ्युदयो नाशो मनोनाशो महोदयः।
ज्ञमनो नाशमभ्येति मनोऽज्ञस्य हि शृङ्खला।।

ness depends on the proper control of the mind, freedom from misery, the drawing of wisdom and the attainment of endless peace, likewise depend on it (Māṇḍukyopaniṣad 3.40)[2]

The mind being very restless, it is very difficult to control. It is as difficult as to tame a wild animal or to control wind. In Bhagvadgita Arjuna says: For the mind is verily restless, O Krisṇa! It is impetuous, strong and difficult to bend. I deem it as hard to curb as the wind.'[3] In Vivekacūdamaṇi, Śaṅkara also emphasizes this view and says: "In the forest-tract of sense - pleasures there prowls a huge tiger called the mind. Let good people who have longing for liberation never go there". (verse No. 176)

Though it is difficult to control it can be controlled with continuous efforts and with the right reasoning. It is possible if one devotes one's entire attention to it and assumes posture after posture, over and over again, unless and until one adopts right reasoning. Even as a wild elephant in rut can not be brought under control by one without the iron hook so also the mind without adopting the right reasoning. Vasiṣṭha has clearly set forth the various reasonings to be adopted by one to bring about the Dissolution of Mind. Whosoever follows them with care will have complete sway over his mind.[4]

Vidyāraṇya states that mind can be controlled in two ways or in other words control is of two kinds: [1] Control by vehemence and [2] control by degrees, i.e. haṭhanigrah and kramanigraha, respectively. In the first the sensory centres are controlled by sheer force where as 'application to spiritual science' knowledge and the rest are the means to be adopted for control of degrees. 'Company of the holy' also proves to be away from vāsanās and control mind. In fact the flow of

---

2. मनसो निग्रहायत्तमभयं सर्व योगिनाम्।
   दुःखक्षयः प्रबोधश्चाप्यक्षयः शान्तिरेव च।।
3. Bhagavdgītā (6.34)
4. L.Y.V., pp. 29-26-27-28

breath (Prāṇa) and strong vāsanā are the two seeds of the tree of the mind. The restraint of these two is the restraint of mind.

As the smith blows upon the live fuel covered with ashes with a pair of bellows, and in consequence of the blast the fire bursts into abright blaze, so does consciousness, covered by ignorance, which stands in the place of the fuel and is the material cause of the mind, on being energized by the flow of breath, quicken itself into innumerable transformations of the mind. From this quickening, namely blazing out of consciousness into what are known as transformations of the mind, arise all ills. This is one phase of the generation of the mind, brought about through the flow of "prāṇa". (J.M.V., p. 90).

To put the same thing in the words of sage Vasiṣṭha:

The flow of prāṇa depend on vāsanā which, in its turn, reacts on the other. When these two stand in the relation of the seed and the sprout to the tree of mind.

He then speaks of the vanishing of both from the vanishing of either:

"Two are the seeds of the mind - flow of breath and latent impressions. On the vanishing of the either, both of them disappear in no time". (L.Y.V., pp. 28-65 & 34)

Thus, are set forth the means of controlling the one and rooting out the other, as well as the results attained therefrom. Hereafter he discusses Aṣṭāṅgayoga of Patañjali, *Vidyāraṇya's* entire chapter on Mano-nāśa indicates his *Yogic Advaita*; the chapter's theme is controlling and restraining the mind by *yogic discipline*; particularly by Patañjali's Aṣṭāṅga Yoga.

Patañjali's Aṣṭaṅgayoga is an important means to dissolution of mind :

As the yoga-bhāsyakāra says : —

योगेन योगो ज्ञातव्यो योगो योगात् प्रवर्तते।

योऽ प्रमत्तस्तु योगेन स योगी रमते चिरम्।।इति।।

i.e., "yoga should be understood by yoga, yoga develops from, yoga: that yogi who is ever vigilant by the practice of yoga finds supreme bliss.

But to be a yogin. one must get mastery on Yama, niyama, āsana, prāṇāyāma, pratyāhāra, dhyāna, dhāraṇā and samādhi.

In the beginning one must be regular in his diet and sleep. A yogin should ever avoid over-eating as well as starvation. As it is said in Bhagavad Gitā: Varily yoga is not for him, who eats too much nor who abstrains to excess, nor for him who is too much addicted to sleep nor even to wakefulness O Arjuna! Yoga kill out all pain, for him who is regulated in eating and amusement, regulated in performing actions regulated in sleeping and working (B.G. 6.16-17). So, Proper diet and Movements ('Yukta āhāra vihāra) are essential.

Yama or forbearance consists in abstaining from killing, falsehood, theft, incontinence and greediness.

(अहिंसासत्यास्तेय ब्रह्मचर्यापरिग्रहा यमाः।
हिंसादिभ्योनिषिद्ध धर्मेभ्यो योगिनं यम यतन्तीति यमाः।।)

Observances are those which turn one away from actions based on desire which are the cause of incarnation and, induce one duly to carry out such disinterested duties as bring about liberation or as quoted in Yogabhāṣya .

शौच संतोषतपः स्वाध्यायेश्वर प्रणिधानानि नियमाः। (2.32)

The result of forbearance and observances are: attainment of supreme vigour (on the confirmation of continence), knowledge of the how and wherefore of existence, i.e. immunity from the fear of rebirth etc. From purity arise disgust for one's own body and dissociation from others. Moreover there arise clear passivity, pleasantness of mind, fixity of attention, subjugation of the senses and fitness for communion with the soul,: Superlative happiness arises from contentment.

Then comes posture. Any posture such as the padma (Lotus) or the swastika or other position of the body which one finds easy and comfortable, as causing no strain or pain in any of the limbs and which keeps the body in a condition so steady as to be free from jerks or tremblings is to him the best posture to be taken. Patañjali concentrates everything about posture in his three sūtras thus:

[1] स्थिरसुखमासनम्। (2.46)

(posture is that which is steady and easy.

[2] प्रयत्नशैथिल्यस्यानन्त समापत्तिभ्याम्। (2.47)

(By slackening of effort and by meditation on Ananta).

[3] ततोद्वन्द्वानभिघातः। (2.48)

(Thence no assaults from the pairs of opposites).

The extraordinary means of attaining it (a comfortable posture) is "meditation on the Ananta", namely such attitude of the *mind* as would take the form of "I am the self same serpant śeṣa or Ananta who stands unmoved, supporting the globe of this earth on his thousand hoods". By this means is generated that subtle attitude which easily procures the requisite posture. Posture being gained, one is no longer assailed by the pairs of opposite', such as 'cold and heat' respect and contempt etc. as before.

Proper place for posture is described in Kaivalya Upaniṣad is thus: One taking a comfortable posture, in some solitary place with extreme (bodily and mental) purity and holding ones neck, head and chest in a line (Kai.5) In a place, even, pure, free from pebbles, fire and sand, as also free from noise and devoid of any great water course, always pleasant and never painful to look at, one should assume ones posture in some cave, sheltered from the blasts (Śve.Up. 2.10).

The diet of a yogin is also important as the posture. A yogin must take moderate food. He should avoid overeating as well as starvation.

Prāṇāyāma comes next to these three i.e. Yama, Niyama and Āsana. Prāṇāyāma is the restraint of breath. Repeated practice of restraint of breath leads to the restraint of mind. As melting separates the pure metal from the one dug up from a mountain and burns up all the dross, so is burnt up all the evil arising out of the improper functioning of the mind, due to the influence of latent impressions, through restraint of breath.[5] So, a yogin should acquire sufficient skill in the practice of Praṇāyāma. One who has acquired such skill has his flow of breath rendered thinner than that of another person. Sruti clearly states: 'having drawn the breath in long enough, he should then give it out very slowly.

Just as a chariot with restive horses is dragged away from its right course to any place whatever and has to be restored to the right path again, by the driver curbing the horses through the reins, so also the mind, dragged hither and thither by the senses and vāsanās is held in perfect check, only when the string of prāṇa is held fast and well (J.M.V., p. 95). Prāṇāyāma the skill to restrain bıeathe is defined thus: 'With prāṇa of proper length, one should thrice repeat the Gāyatri, the Vyāhṛti, the Praṇava and the Śiras - this is praṇāyāma. prānāyāmas are said to be three in numbers, recaka, pūraka and kumbhaka. Drawing the air up, completely exhausting the air from the ākāsa (space) and creating a void therein . one should control the void thus creates: This is 'recaka'. As one sucks in water through the stalk of a lotus, so should one draw air in (for prāṇayāma) this is pūraka Neither inhaling nor exhaling any air, nor moving any of one's limbs, one should direct oneself towards prāṇāyāma: this is 'Kumbhaka' (Anṛopaniṣad 11 14)

By the constant practice of praṇāyāma and adopting the method expounded by the teacher the mind of a yogin becomes fit for the stages of contemplation,

धारणासु च योग्यता मनसः। (2.53)

Patañjali sets forth the result of prāṇāyāma thus:

ततः क्षीयते प्रकाशावरणम्।

The light i.e. the light of sattva that which obscures such

5. Amritanadopanisad 7
यथा पर्वतधातूनां दह्यन्ते धमनान्मलाः।
तथोन्द्रियकृता दोषा दह्यन्ते प्राणनिग्रहात्।।

light, is tamas, the cause of sleep sloth and the like this is removed by the practice of pranayama.

Next to prāṇāyāma is pratyāhāra, i.e Abstraction When five senses and the forth mind are turned off from their respective objects, they should be thought of at so many reins of the inner self it is called pratyāhāra. The result of abstraction is given in the aphorism. ततः परमा वश्यतेन्द्रियाणां।[6] Then follows the greatest mastery over the senses.

Contemplation or dhāraṇā is the fixing of the mind on something that state wherein the wise man considers the mind as the ever active instrument of imagination, projects it on the self and retains it in the self. Same condition is described as dhāraṇā (contemplation)[7]. Let the mind which thinks of every possible object, think of itself and of nothing else' effort in this direction is called the fixing of the mind on itself.

This leads to Dhyāna. The unity of the mind with it, is absorption or dhyāna. (3.2). It means the flow of mind should be in one direction of "tattva', to put in other words of Sarvānubhava: Knowledge of the kind just described comes of the one pointedness of the mind. The method to be employed for acquiring the means to such knowledge, to wit, absorption, is as follows : One should meditate upon the residual sat - cit and Ānanda, after causing the several orders of creation to melt away, by reversing the process of evolution in his mind. That stream of the transformtions of mind ever conscious of Brahman and devoid of least vestige of egoism, is the result of confirmed practice of dhyāna. It is also known as Samprajñāta Samādhi or concentration of the conscious variety.

Thus, a yogin reaches at the stage of samadhi a last stage of aṣtānga yoga.

तदेवार्थमात्र निर्भासं स्वरुपशून्यमिव समाधिः। (3. 3)

---

6. Amritanadopanisad 2.55
7. Ibid. 3.15

The same, when conscious only of the object, as if unconscious of itself is concentration. Contemplation, absorption and concentration are supposed to be the internal auxiliaries of conscious Concentrations as they relate to the mind. where as the first five are supposed as the external, Patañjali has described Samprajñāta and Savikalpa kinds of concentration: but they aim at certain occult powers such as being invisible etc not at the absolute concentration i.e. liberation. It is Nirvikalpa-samādhi or absolute concentration alone which leads to liberation. Absolute interception or Nirodha samadhi is the last stage of samādhi.

It means the utter emptying of the mind of all thoughts other than the one on which it stands intercepted, so to speak The supreme fullness of mind with this idea, even up to the forgetfulness of such experience taking the positive aspect of the state also can be described as Absolute interception' . Such absolute interception is gained only after strong personal efforts. The ultimate tranquility is gained stage by stage.

The wise man should reduce all speech to thought, all thought to the self that discriminates (buddhi), the discriminating self into the great self (mahat), and the great self into the self that is all peace and tranquility (Kathopanisad 1.3.13).

So the control of speech also is necessary. When control of speech gets confirmed into a habit and speechlessness becomes part of our nature, then the thoughts are reduced to the self that discriminates.

Thus all these factors help to dissolve the mind. Though its not easy to curb the mind it is not impossible also. Lord Krisṇa says: "Without doubt, O mighty armed! the mind is hard to curb; but it may be curbed by constant practice and by dispassion. Yoga is hard to attain, me thinks, by a self that is uncontrolled, but by the self controlled it is attainable by properly directed means (Bhagavadgītā 6.35.36).

असंशयं महाबाहो मनो दुर्निग्रहं चलम्।
अभ्यासेन तु कौन्तेय वैराग्येण च गृह्यते।।

It is only 'constant practice' and 'dispassion' which helps to control the mind.

असंशयतात्मना योगो दुष्प्राप इति मे मतिः।
वश्यात्मना तु यतता शक्योऽवाप्तुमुपायतः।।

Vidyāraṇya quotes the example of tittika birds from Loghu Yoga Vāsiṣṭa (3.41) which reveals that Lord blesses the yogin, who applies himself with untiring efforts to the highest dharma, namely the control of mind. The fable well versed in Laghu Yoga Vāsiṣṭa is like thus: The Ocean carried away with its receeding tide the eggs of some birds, laid somewhere on its beach. (Enraged at this), the bird resolved upon drying up ocean, and begain to empty out the water drop by drop with its bill, Though dissuaded by other birds of its kind, nothing daunted the bird sought their alliance. The sage Narada observing the trouble and worry of so many birds constantly flying to and from the ocean, took compassison on them and sent Garuda, the Lord of birds to go and assist his kindled. Thereupon, the ocean which began to dry up by the flapping of Garuda's winds, was filled with consternation and restored the egges of puny birds. Thus the untiring efforts lead to favourable result continuous efforts lead to dissolution of mind.

If one thinks to be a yogi. how and in what way should a novice fix his schedule of a mind? This also is described in Laghu Yoga Vāsiṣṭa where Vasiṣtha says:

"The right course to be adopted by one who is in his novitiate is this: Two parts of mind must be filled with objects of enjoyment, one with philosophy and the remaining one with devotion to teacher.

'Having advanced a little, he should fill one part of the mind with objects of enjoyment, two parts with devotion to the teacher and the remaining one with getting an insight into the meaning of philosophy.

When he has attained proficiency he should everyday fill two parts of his mind with philosophy and supreme renunciation and the remaining two with mediation and devoted service to the Guru (L.Y.V. 21/36.37.38)

By objects of enjoyment here means going about for alms, to keep the body and soul together and such duties as pertain to the varṇa and āśrama to which one belongs. Step by step a yogi attains the mastery over all the eight aṅgās of yoga.

A student, who launches upon a course of the study of the śāstras, stand in need of help in the form of interpretation and explanation of every line to him, till he has gained proficiency, but as soon as he has gained it, the rest of the book is plain to him without such help. In the same manner, the yogin, who has mastered thoroughly every preceeding stage finds himself the means to accomplish the stage that succeed. The science of yoga, thus concerns itself with concentration as the proper method of the treatment of mind.

Vidyāraṇya clearly indicates that through purified by samadhi, the tvaṁ-padārtha or the substratum of ego is realised. Yoga leads to the stage of final realization, Yoga helps to still the mind by samadhi. But is 'Brahmavidyā' or the knowledge alone which has to be induced by the Guru through the (mahāvakya 'Tat tvam asi') helps to reach the stage of realization. Such realization is the goal of human life and not the samadhi. So neither the yoga alone nor the knowledge alone lead to the supreme bliss, but both are equally important. As Vasiṣtha preaches: O Rāghava! Yoga and knowledge are the two paths leading to the Dissolution of Mind; yoga consists in interpreting transformation of the minds and knowledge in the proper viewing of things. To some, yoga, is unattainable while to others is denied the capacity to judge aright hence the Supreme Lord prescribed these two paths. (L.Y.V. 27.72 / 29-60)

Vidyāraṇya notes that some yogins attain supernatural powers (siddhis) like flying which are not seen in Jīvanmuktas. However, one ignorant of the self can get powers like flying and such powers are not the object of knowledge of self. Mere persuit of powers indicate one's ignorance and does not help one to attain the highest end. These desires also must cease as these desiring powers are ultimately the obstacles. Again since such wonderful powers come from the self, any one who knows the self could achieve them. This passage echoes that Brahman knowers can certainly gain and use supernatural powers but they are detached to do, so such claims affirm the ultimacy of the path of knowledge over yogic mastery.

Throughout the chapter on extinguishing the mind, Vidyāraṇya examines various types of samādhis which aim to still the mind; such as Asamprajñāta - samādhi, Nirvikalpa - samādhi, Nirodha samādhi, samprajñāta, Savikalpa samadhi etc. which aim to still the mind; and urges the cessation of the activity through yogic practices. However, he says even the highest (nirodha or Aasamprajñāta) yogic samadhi is not the final goal. It leads to knowledge of the self. Here he indicates the limitations of Yoga. Texts of yoga school focus only on stilling the mind by samādhi, they do not explicitly mention the Advaitin realization of the self (i.e. आत्मदर्शन) and seeing the self is the ultimate goal. He claims that even after one attains the highest samādhi. another mental manifestation (vṛtti) called Brahma-vidyā must arise in order to reach identity with Brahman. This vṛtti appears by heaping the Upaniṣadic goal statement (mahāvākya) like 'tat tvam asi' (thou art that) by a guru.[8]

So the yogic practices as well as knowledge lead to Jīvanmukti. Lassitude, distraction and taintedness are cured in the state of Brahman, and evenness is attained. 'The supreme Lord pervading all things is evenness itself. (B.G. 13 -28). The supreme bliss which the reason can grasp, is beyond senses (B.G. 6 -21). The bliss experienced by the mind, purified of all dross from the practice of concentration and fixed in the Ātman, can not be adequately described in words but can be comprehended only by the inner sense acting by itself. (Maitreyi Up. 4.9).

Thus, Vidyāraṇya advices one to be a yogin to attain

8. J.M.V., p. 115

Jīvanmukti. 'A yogin', whose mind is free from lassitude, distraction, taintedness and taste for pleasure, becomes undisturbedly fixed in Brahman.

लयविक्षेपकषायेषु परिहृतेषु परिशेषाच्चितेन समं ब्रह्म प्राप्यते।

This is called the highest goal, wherein all the five senses and the mind remain in full control, and wherein even the intellect does not function. This steadying of the senses is called yoga the yogin is wide awake in that condition; for yoga is evolution coupled with involution'. (Kaṭhopaniṣad - 2.3-10-11).

Yoga if neglected, is the cause of evolution of activity in the senses; and if properly practised, it leads to the involution of such activity. 'Yoga is the supression of the transformations of the thinking principle (1.2).

Yoga should be practised for a long time, measured in years, nay in births.

अनेकजन्म संसिद्धस्ततो याति परां गतिम्। (B.G. 6.45)

Fully perfected through manifold births, he reaches the supreme goal. So yoga should be practised for a long time, without intermission, with earnest devotion. His concentration should not be shaken by any heavy sorrow, absolutely detached from the worldly pleasures and pains, he should have mastery over his mind in cessation of all desires. Such a Jīvanmukta is forever freed from all incarnations.

Lord Krişṇa says:

तपस्विभ्योऽधिको योगी ज्ञानिभ्योऽपि मतोऽधिकः।
कर्मिभ्यस्ताधिको योगी तस्माद्योगी भवार्जुन।।

' i.e. The yogin is a greater than a ascetic, he is thought to be greater than even knowers; the yogin is greater than the men of action; therefore, O Arjuna, become a yogin.'

Such is the importance of being a yogin. In this way, Vidyāraṇya has explained the two main paths. Vāsanākṣaya and Manonāśa which with the help of knowledge lead to Jīvanmukti.

Chapter - X

THE PURPOSE OF THE ATTAINMENT OF JĪVANMUKTI

Vidyāraṇya gives a consistent and logical justification as to why one strives for the attainment of liberation-in-life. In the fourth chapter of Jīvanmuktiviveka he enumerates five purposes of the attainment or Jīvanmukti. They are— ज्ञान रक्षा तपो विसंवाद-भाव दुःखनाश सुखाविर्भावाः सन्ति पञ्च प्रयोजनानि।

i.e. [1] Preservation of knowledge

[2] Practice of penance

[3] Absence of discard

[4] Cessation of pain and misery

[5] Attainment of Supreme Bliss.

He then discusses and evaluates them, thus :

The first aim of the attainment of Jīvanmukti is the preservation of knowledge (jñānarakṣā). In fact the knowledge has already dawned in the enlightened. It is fully safeguarded when one practices dissolution of the mind and obliteration of latent desire But Vidyāraṇya is of the opinion that doubt and false knowledge may crop up in the mind, even if knowledge has been dawned. He says :

तस्माद् विदितेऽपि तत्त्वे विश्रान्तिरहितस्य शुकराघवयोः इव संशय उत्पद्यते। स च अज्ञानमिव मोक्षस्य प्रतिबन्धकः॥

In Bhagavadgitā Lord also says 'But the ignorant, faithless, doubting self goes to destruction. neither this world nor that beyond, nor happiness is there for the doubting self (4. 40).

अज्ञश्चाश्रद्दधानश्च संशयात्मा विनश्यति।
नायं लोकोऽस्ति न परो न सुखं संशयात्मनः॥

By faithlessness is meant false knowledge whereas ignorance and false knowledge stand in the way of liberation alone, doubt prevents both liberation and the enjoyment of worldly pleasures; for doubt always hangs as it were, between two opposite extremes. Whenever there is a tendency for worldly enjoyment, then the mind, yearning for liberation, curbs such tendency; and vice versa. Whenever there is the yearning for liberation the mind, tending towards worldly enjoyment, curbs such yearning. Hence, one desirous of liberation should, by all means, get oneself rid of all doubts, for the reason that there is no happiness whatever for the sceptic (J.M.V. page no. 132)

The instance of false knowledge is seen in case of 'Nidāgha' who being once Enlightened by the great sage Ṛbhu, was swayed by false knowledge which was later on overcome by attaining proper peace of mind. Vidyāraṇya, therefore insists that there is the supreme necessity even for those who have realised the truth, to resort the unconscious variety of Trance, the Dissolution of the Mind, with a view to destroy mental suffering like doubt and false knowledge with the same in view. The Lord in the Gītā says to Arjuna:

यततो ह्यपि कौन्तेय पुरुषस्य विपाश्चितः।
इन्द्रियाणि प्रमाथीनि हरन्ति प्रसभं मनः॥ (2.60)

(O son of Kunti! the excited senses of even a wise man, though he be striving, impetuously carry away his mind)

Vidyāraṇya makes clear that one may ask: Why the saving knowledge should be safe-guarded at all ? i.e. it appears to be a mutual contradiction to say. 'A liberated reck not' and 'they conduct themselves (in their everyday life). There is no such contradiction involved; for both these positions could be reconciled by taking into consideration the degrees of the knowers of Brahman viz. [1] The knower of Brahman (Brahmavid) [2] Brahma-vid-vara ie. knower of Brahman of higher type [3] Brahmavidvarīya i.e. the knower of Brahman of next higher type and last [4] Brahmavidvariṣta i.e. knower of Brahman of highest type (J.M.V., p. 134) So these are the degrees of knowers of Brahman. They should be understood to belong respectively to the four stages, beginning from the fourth of the seven stages of knowledge. These seven stages are thus described by Vasiṣṭha:

ज्ञानभूमिः शुभेच्छाख्या प्रथमा समुदाहृता।
विचारणा व्दितिया स्यात् तृतिया तनुमानसा।।
सत्त्वापत्तिश्चतुर्थी स्यात् ततोऽसंसक्तिनामिका।
पदार्थाभाविनी षष्ठी सप्तमी तुर्यगा स्मृता।।

(The first stage of knowledge is known as 'ardour', the second is the 'spirit of enquiry', the third 'attention', the fourth 'attainment of sattva (purity)', the fifth 'indifference', sixth 'Oblivion', and the seventh 'transcendence' (of the previous three).

At the first stage one, takes the help of philosophy and the company of wise. The spirit of enquiry consists of that constant application of the mind in search of that. Constant application of the mind in search of what is wholesome, born of the study of philosophy the company of the wise and the repeated development of detachment. These two stages lead to the third 'attenuation'! Mind, becalmed of all that belongs to the objective sphere, finds complete rest in the pure bliss of sattva. The first three stages are only the means of knowledge and can therefore not be included in the Brahma-vidya proper, for in them the sense of pseudo-reality is not removed The first three are therefore assigned to working condition whereas the fourth stage is called the "Brahmavid'.

The last three stages are only the sub-divisions of 'Jīvanmukti'. They are due to the difference in the degrees of restfulness arising from the constant practice of unconscious variety of concentration. The yogin in the fifth stage may, of his own accord, issue out of the said concentration. Such a yogin is called the Brahmavid-vara (belonging to the first degree) In the sixth stage, he issues out of his concentration only when aroused by others standing beside him and is known as Brahmavid-variya (belonging to the higher degree). The ascetic in the seventh stage known as "transcendence' knows no issuing out (of his ecstatic concentration), either of himself or from any other cause. A yogin is Brahmavidvariṣtha here. Thus mutual conflict apparantly involved in the two statements 'whenever their attention is drawn to it. by persons standing by, and the liberated recks not' does not arise, as they relate to two different stages.

The three stages beginning with the fifth, comprised in Jīvanmukti, having been realised, as there is no possibility of the occurrence of doubt and false knowledge on account of the appearance of duality in those conditions, the resulting knowledge is fully safeguarded from all impediments whatever thereafter. So then, this safeguarding of knowledge is the first aim to be sought in the attainment of Jīvanmukti.

Tapas :

The second aim is the practice of penance. "The stages of yoga should be looked upon as making up the practice of tapas as they are the causes that contribute to the exaltation of the ascetic to the condition of the Gods and other higher beings. As the Śrutis says: By penance did the God of yore rise to the condition of Gods, by penance again the ṛṣis gain heaven (Tai. Bra 3.12. 3). The highest penance consists in the one pointedness of the mind and the senses; that is better than all religious practices, that is the highest dharma, says smṛti. Though, according to this principle, no highly incarnation is attainable through penance, still it has a distinct purpose to serve, namely, the good of the world.

The World to be served according to Vidyāraṇya is divided

into three kinds 1. The world of Pupils; 2. that of devotees and of 2. Neutrals (i.e. neither the one nor the other). Of these the first, in virtue of the highest faith in the truthfulness of the Guru, who is a real yogi ever centered within, has implicit trust and confidence in the truth taught by him and attains mental composer in a very short time. Smriti says. "He who is absolutely devoted to God and verily looks upon his Guru as his God - to that great soul are revealed the sublime truths of what is imparted (by the treater).[1] The 'devotees' acquire, as it were the penance practised by the yogins, merely by rendering service to him by way of accommodating him in the matter of food, habitation and the like. Neutrals follow the rectitude path of the yogins and they get rid of sins being blessed by the glances of the yogins.

It is not merely the activity of yogin. such is sanctioned by religious works, that constitutes penance, but all worldly activities of his are also of that nature. Chāṅdogya upaniṣad, describes penance thus :

'अथ यत्तपोदानमार्जवमहिंसा सत्यवचनमिति ता अस्य दक्षिणाः।' ।।इति।।

(Now what are the yogin's penance ? Charity, straight forwardness, nonviolence truthfulness,—these are the gifts of sacrifice) (Ch. Up. 3.17.4)

A yogin's whole life is a sacrifice, attended with the giving as gifts, of all that he has. He who holds the life career of a yogin, his life is identical with all the rituals spoken of in the Vedas commencing from 'Agnihotra' and ending with the saṅvatsara-sattra, and becomes one with the sun and the moon in proportion to the strength of his faith. If his faith is less fervid, he only attains the solar or lunar sphere and having there shared the divine splendour of either the sun or the moon, he rises further to satyaloka and acquires the glory of the four faced Brahmā". Having acquired knowledge while in that sphere, he thence gains the glory of Brahman the entity which

1. Śvetāśvatara Up. 6.23

is all being, intelligence and bliss. In other words, he gains kaivalya. Thus, is established the second aim of Jīvanmukti namely, the Practice of tapas."[2]

The third aim of 'Jīvanmukti' is the absence of discard. The foremost among yogins, being prone to introspection and ever detached from all external intercourse, no one whether worldly minded or doctrinarian, has ever any cause for conflict with him. This conflict, in the case of worldly minded, takes the two forms of dispute and censure. The yogin being ever devoid of anger and the rest, how can the worldly minded ever find cause for dispute with him? asks Vidyāraṇya. He says 'A yogin should return not anger for anger, but wish him well, who threats menaces him with evil, should cooly treat all tall talk (tending to provoke him) and by all means displease no one.[3]

A yogin looks even upon a disputant as (identical with) himself, where is the room for his gloating over his victory ? He does not care to establish the position of the philosophy he follows. Nor is it probable that cultured believers of whatever school, bearing agnostics, having faith in the reality of liberation will ever take exception to the yogin and his ways. Any school of thought which believe in mokṣa. differ in their method of exposition, they are unanimous in respect of recognising yama, niyama and other stages of yoga, as the means to liberation. As Vasiṣṭha says : O best of intellectuals! in him readily seek shelter all spotless systems of knowledge, whose present incarnation is that last and the best, even as pearls seek shelter in the best bamboo. Nobility, affability, friendliness, sweet temper, detachment and high attainments - all ever resort to him. All men seek him who is sweet with pleasing conduct, even as the wild deer in the forest seek after the melodious note of the flute. He is the liberated one, who though, on account of the supreme tranquility of his mind, he remains, as it were in kind of sleep is ever mentally wide awoke and is ever sought after by

---

2. J.M.V., pp. 142-143
3. Mansumṛti 6.48. 47

the wise, on account of his proficiency in all learning, even as the moon full of all the kalas. is by the Gods[4] . Thus is clearly made out the third aim of Jīvanmukti— The absence of discard.

The fourth and the fifth objects are dukkhanāsa and Sukhāvirbhāvah i.e. cessation of pain and misery and the genesis of Bliss. The Vedic texts, such as Alas! why I have done no good, why I have committed sin! These do not cause pain to this one'[5] be speak the entire destruction of pain taking the form of regret relating to virtuous and vicious conduct. These have a bearing on the next world.

The genesis of supreme bliss' can be attained by the attainment of all desires, by discharging of duties (of every kind) enjoined on one' and 'the achievement of the end and aim of life. He who realises 'I am only that Brahman, which is all consciousness and the witness in the bodies of all beings whatever, from Hiranyagarbha down to the lowest order of animate creation, is necessarily the witness of the bodies of all. even as he is the witness of his own body. The knower of Brahman being unaffected by desire for the enjoyment of things enjoyed by the world means the attainment of all objects of desire[6].

'The discharging of duties (of every kind) enjoined on one is thus set forth in the third Chapter of Bhagavadgita:

Lord Krishna says :

ज्ञानमृतेन तृप्तस्य कृतकृत्यस्य योगिनः।
नैवास्ति किंचित् कर्तव्यमस्ति चेन्न स तत्त्ववित्।।
यस्त्वात्मरतिरेव स्यादात्म तृप्तश्च मानवः।
आत्मन्येव च संतुष्टस्तस्य कार्य न विद्यते।। (3.17)

("Revelling in the nector of knowledge to his hearts content,

4. L.Y.V. 19. 9-11
5. Taitt. Up. 2.p
6. J.M.V. (p.164)

fully satisfied with the complete discharge of every duty, there remains nothing for the yogin to do; if aught remains, then he can not be reckoned as having attained knowledge.

"But the man who rejoices in the self with the self is satisfied, and is content in the self, for him verily nothing to do".

The 'achievement of the end and aim of life' is also touched upon in *śrutis*

अभयं वै जनक प्राप्तोऽसि। (Br. Up. 3.2.2)

(thou, O janaka! has realised entire fearlessness)

तस्मात्तत्सर्वमभवत्। Br. Up. 3.4.40)

(Hence he became all that)

ब्रह्मविद् ब्रह्मैव भवति। Mun. Up. 3.2.9)

(The knower of the Brahman becomes Brahman alone)

Thus Vidyāraṇya puts forth five purposes of Jīvanmukti. After attaining Jīvanmukti one's knowledge is safeguarded. His whole life becomes a practice of penance. All sufferings of such a man come to an end and he realises the supreme bliss. His bliss is superior than the pleasure on earth and heaven. Nothing remains to be attained after attaining 'Jīvanmukti'.

Chapter - XI

A PARAMAHAṀSA YOGIN

The last chapter of 'Jīvanmuktiviveka' deals with the renunciation of Paramahaṁsa Yogin i.e. Vidvat-saṅyāsa Vidyāraṇya recounts Jīvanmukta s way of life most clearly in the final chapter of his work. It forms a commentry on the Paramahaṁsa upaniṣad. He describes how the supremely ascetic vogin or realised renuncient, (Paramahaṁsa or vidvat-saṅyāsin) interweaves yogic practice and renunciation. The discussion illustrates both the path to and life styles in living liberation. Let us see the discussion in detail:

In the beginning of the chapter Vidyārarnya differentiates carefully the difference between a mere yogin and a paramahaṁsa yogin. The mere yogin, having no climps of the essence, attaches himself to the use of clairvoyance, floating in the air and other occult powers of the kind. He puts forth his effort by employing various means, such as saṁyama (restraint)saṅyāasa in the direction of the acquisition of such powers. Thus, he turns aside from the path leading to the real goal of existence. These are obstacles in the way of Samādhi and are powers in moments of suspension.[1]

On the other hand, the mere paramahaṁsa, in view of his fully realised the essence, comes to the conclusion that such occult powers are mere trash and treats them with indifference.

---

1. J.M.V. 4/9

The curiosity of the Paramahaṁsa is not roused by all this wondrous panorama, as he knows fully well that these powers of the self that is all consciousness manifest themselves in diverse ways. In Paramahamsa upaniṣad. (I), Lord Brahma replies to Nārada: That the path of Parmahaṁsa is the most uncommon path and is very rarely to be met with in this world though the path is difficult to follow, its not impossible. One should not remain without striving towards the end, hardly anyone can reach the end. Lord says, 'even if there be only one such sage, he alone rests always in the ever holy, he alone is the Vedapuruṣa.' Thus say those who are in the knows. In Bhagavadgitā also Lord Kriṣhna says :

मनुष्याणां सहस्त्रेषु कश्चिद्यतति सिद्धये।
यततामपि सिद्धानां कश्चिन्मांवेत्ति तत्त्वतः।। (7.3)

(Among thousands of men, scarce one strives for perfection, of the successful strivers, scarce one knows Me in truth)

Vidvāraṇya, here explains : Should there be found any where any single yogin Paramahaṁsa who is in accord with the principle laid down in this verse of the Bhagvad Gitā, then, such is verily the one who rests always in the ever holy, i.e. paramātman. The mere Paramahaṁsa, though knowing, runs after the objective sphere, owing to the lack of mental tranquility and rests not in Brahman. Only the Veda-Puruṣa, i.e. The Supreme Soul established by the Veda, the persons well-versed in that system of knowledge, which deals with the realization of Brahman and the mode of attaining mental tranquility. All men regard the yogin-paramahaṁsa as one who takes his stand on Brahman, but the aforesaid men in the know would not be content even with this but would look upon him as in the state of Brahman itseif. The smriti says:

दर्शनादर्शन हित्वा स्वयं केवल रूपतः।
यस्तिष्ठति स तु ब्रह्मन् ब्रह्म न ब्रह्मविद् स्वयम्।।

(He who stands centered entirely in himself, taking no heed of perception or non-perception, is O Brāhmaṇa! Not simply a knower of Brahman but Brahman itself)

'महापुरुषो यच्चित्तं तत्सर्वदा। मय्येवावस्थापयत्ति, तस्मादहं च तस्मिन्नेवावस्थित:।।' इति।।

(The Mahāpuruṣa concentrates his mind on Me and in consequence I also take my stand (always) in him).

This is the status of a yogin Paramahaṁsa. He is entitled as a great soul (mahapuruṣa, a mahātman).[2]

Here we come to know that Vidyāranva differenciates a yogin Paramahaṁsa from a mere yogin and a mere paramahaṁsa The word Paramahaṁsa is a compound word parama + haṁsa (Parama is great and haṁsa is realized soul) which means an ascetic of the highest order; one who has controlled and subdued all his senses by abstract mediation'.[3] Śaṅkara defines the term as Great knower of the self; Parama means Great and hamsa means the knower of the self. Vidyāraṇya also uses the term in the same sense. Such a yogin Paramahaṁsa always keeps his mind absorbed in Supreme Self. All the transformations of his mind connected with the affairs of the world, in his case, are entirely suspended by practice and detachment'.

Such a yogin-paramahaṁsa should renounce his wife, children, relatives, friends and the rest along with the tuft (on his head), the sacred thread and the study of the Vedas...... in fact all ceremonial worship of every discription nay. leave worldly life in its entirety and should betake himself to a mere lion-hide, a bamboo-stick and a blanket, alike with a view to providing for his bodily comfort and doing good to the world at large'. This is the path of a yogin-paramahaṁsa.[4]

In the above lines, the words 'and tne rest' put after wife, children, relatives, friends refer to servants, domestic animals, houses, fields and all other temporal goods. The words 'study

2. J.M.V., pp. 150-152
3. Advaitic Concept of J.M.: L.K.L. Shrivastava
4. J.M.V., p. 153

of the Vedas' refer to giving up of even those aids to study which are of use in understanding the correct import of the Veda as grammer. metaphysics, yogic and those that amplify the Veda, as epics, the Puraṇas etc. Even the pieces of lion-hide, bamboo-stick, blanket and a pair of wooden sandais are not absolutely (तच्च न मुख्योऽस्ति।) necessary. (Parama. Up. 1)[5]

Just as a Paramahaṁsa in the stage of the renunciation of the seeker has as his essential condition the absence of the tuft and the sacred thread, so also the yogin paramahaṁsa should have the absence of the stick and the blanket as his essential conditon. It is because his mind may not get distracted with the thoughts of acquisition or preservation of these things. Yoga that consists in supression of the transformation of the thinking principle', will be unattainable to him. Then how should he manage without these things is suggested in Paramahaṁsa Upaniṣad. 2.

न शीतं न चोष्णं न सुखं न दुःखं।
न मानावमानौ च षडूर्मिवर्जम् ।।इति।।

[i.e. neither cold nor heat, neither pleasure nor pain, neither courtesy nor contumely (affects him). Also he is devoid of all six urmis (conditions)]

As a child, absorbed in play, feels no cold or heat, so also a yogin paramahaṁsa engrossed in the supreme self knows no cold whatever. His mind is entirely rid of all transformations of any kind. The six conditions (urmis) are hunger and thirst, grief and delusion, dotage and death. These three pairs are the characteristic qualities of the vital breath, the mind and the body, but a yogin paramahaṁsa is aloof from all these. He is also aloof from many other bad qualities as it iṣ said:

'निंदागर्वमत्सरदम्भदर्पेच्छाव्देष-सुख-दुःख-कामक्रोध-लोभमोहहर्षा-सूयाहंकारदीश्च हित्वा।' ।।इति।। (परम. 34.2)

He (stands) unconcerned with slander, pride, rivalry,

5. Ibid. p. 156

ostentation, browbeating, covetousness, vindictiveness, opulent ease, chili penury, anger, niggardliness, peevishness joy, envy, egoism and the like.

One's body is looked upon as a corpse, as it were, as the same has been disvalued (Paramahaṁsa Upaniṣad).

संशयविपरीतमिथ्याज्ञानानां यो हेतुस्तेन नित्यनिवृत्तः।।इति।।[6]

He is constantly free from all these and permanently enlightened in it. The yogin who is ever awake in the light of Supreme Ātman is said to be permanently enlightened in it. It is only the Yogin Paramahaṁsa who bearing the text. The Brāhmaṇa with firm resolve, having known it, should ever continue in the knowledge of the self (Br. 4.4.21), gets over all mental distractions, with the force of yoga and continues in the knowledge of the self without intermission. Thus this enlightenment is everlasting.[7] He gains conviction of the truth that is the theme of all Upanisads, is his own self and none other than his own self abides in it.

Vidyāraṇya also discusses the manner in which the realization of Brahman is attained by the Yogin Paramahaṁsa. He experiences :

तं शान्तं अचलं अव्दयानन्दविज्ञानघन एवास्मि।
तदेव मम परमं धाम। (परम. उप. 2)

(I am only he. who is tranquil and immobile, that non-dual essence of consciousness and bliss: That alone is my highest status).

The yogin paramahaṁsa is the one belonging to the fourth order spoken of in the Veda; he is possessed of forbearance and observance; he is Viṣnu incarnate; and the bearer of the three *daṅdas*.'

6. J.M.V., p. 159
7. J.M.V., p. 160

The three daṇḍas are control of speech; of mind; and of action. Control of action consists in moderation in diet. These marks of a tridaṇḍins have also to be observed by the paramahaṁsa yogin.[8] The paramahaṁsa yoain may eat the food once or twice (as is necessary), but he should by all means, constantly keep up the flow of knowledge within. Knowledge is supposed to be the superior kind of daṇḍa for a paramahaṁsa yogin.

Clad only in space (i.e a vacant space around) as his covering, not obliged to salute (others), nor take part in rituals for propitiating the manes of the departed ancestors, affected by neither slander nor praise, the mendicant should keep an open mind (adopt an attitude of self effacement). No invocation, nor sending away, no incantation, nor meditation nor devotion, neither connotation nor denotation, neither the sense of seperation nor of identity, I you (न च अहं न त्वं), neither all, nor (the rest) the mendicant has not even a residential abode. He should not at all (gather round him a set of disciples, nor even have a look at them. (Parama. Up. 4). A paramahaṁsa yogin goes beyond everything. He reaches the delightful condition of fulfilling all duties on realising at all times 'I am that Brahman,' described in Upanisads as that Supreme Self which is all Bliss and the one Consciousness such a paramahamsa yogin becomes fully satisfied.

ज्ञानामृतेन तृप्तस्य कृतकृत्यस्य योगिनः।
नैवास्ति किंचित् कर्तव्यमस्ति चेन्न स तत्ववित्।।[9]

Revealing in the nector of knowledge to his hearts content, fully satisfied in the complete discharge of every duty, there remains nothing for a yogin to do, if ought remains, then he can not be held to have attained knowledge. Thus Vidyāraṇya has discussed the path, the status and conduct of a yogin paramahaṁsa i e a Jīvanmukta. He is rightly called 'liberated while living.' The toughest path yet it is full of satisfaction.

8. J.M.V., p. 164
9. J.M.V., p. 173

CONCLUSION

All the Indian Philosophical Systems agree in having the attainment of Moksa as the goal of life and knowledge. Knowledge is sought for attaining complete freedom from the suffering of the worldly life and from the cycle of birth and death by destroying Adṛṣṭa forever.

The Vedic idea of Mokṣa consists in the termination of pain and sufferings of the worldly life and in the attainment of a more happy life here on the earth, as well as in the heaven. It also consists in the attainment of immortatity and in enjoying divine happiness in the company of Gods. The Vedas hold that it can be attained by sincere prayer and sacrifices.

The Upaniṣads depict their Mokṣa as the attainment of immortality by means of self realization. It is an experience of identity of the soul with Supreme Self. The soul experiences a kind of supersensuous happiness by overcoming the limitations and imperfections of Saṁsāra. It transcends duality and relativity. Though Upaniṣads do not mention clearly and seperately the term 'Jīvanmukti'. The charactaristics of Jīvanmukta are clearly stated in it and it is also said that 'Brahman can be known here on earth while embodied.

It is Śankara. who emphatically hold that one who is free from all attachments, desires, from nescience, karma, finitude who has experienced Supreme Self, he is a Jīvanmukta, living God on earth. For Śaṅkara Mokṣa consists in actually experiencing the Brahman, where all distinctions, limitations, dualities, relativity melt away This can be done by the knowledge of the self, which is Brahman, by means of purification of mind, by self control and yogic practices Gauḍapāda does not use the term 'Jīvanmukta' or 'Jīvanmukti', though he supports Jīvanmukti, but he uses the terms such as 'muni', 'muninetara' (saint, super-saint), 'dhir' 'tattvavid' (knower of the Reality) 'mukta' (liberated), 'Viśārada' (free from fear), sanātana (ever the same) to denote a Jīvanmukta. Dukha-Kṣaya (destruction of misery), state of Supreme happiness (Uttama-Sukha) abhaya (where there is no fear) etc. as the state of 'Jīvanmukti.'

Post Śaṅkara Advaitins also accept Jīvanmukti but their approach towards it differs. They accept the concept in their own way. Maṅdan Misra, In his Brahma-Siddhi, he takes sthitaprajña of Gītā as still a sādhaka and not a Siddha, who has totally annihilated Avidyā. For with the total annihilation of nescience the body must necessarily fall off.

Sureśvara followed the views of Śaṅkara. The views of Sarvajñātman Muni resemble with those of Śaṅkara and Sureśvara but he does not give full support to the concept of Jīvanmukti. Like Maṅdana, he also docs not think the possibility of the concept on logical ground; because according to him, neither Avidyā nor its products, for example the body, can exist after the down of knowledge. It is all destroyed in its fire. Thus the only mukti acceptable on this ground is. sadyomukti (i.e. immediate deliverance). All other Advaitins like Vācaspati Miśra, Prakāśātman. Ānaṅdānubhava, Chitsukhācārya, Ānaṅdagiri all accept Jīvanmukti. Vidyāraṇya not only accepts and supports Jīvanmukti, but proves it to be a unique concept which leads to the ultimate goal of human life, which teaches a person how to live happily. Giving the proofs from Bhagavad-gītā. Lagu Yoga-Vāsiṣṭha, Paramahaṁsopaniṣ ad, Māṅḍukyopaniṣad and many other Upaniṣads and religious works, he proves that a person can get here on earth pure happiness i.e. happiness unmixed with pain. A man can remove his sorrows and there is a path named 'Jīvanmukti' which leads to the cessation of sorrow. Though jīvas in its saṁśāri state suffer various kinds of afflictions, though, it experiences pleasure as well as pain, it is Jīvanmukti which yields eternal peace and satisfaction to a saṅsāri jīva. Jīvanmukti is rightly called as living liberation. It is a path of Karma-Saṅyāsa or the path of Karma which are performed for the welfare of mankind, and the maintainance of world order and active participation in removing the sufferings of others. He is called a Jīvanmukta who has renounced all his desires passions wealth, property, fame, prestige, who performs selfless activities for others' benefit. He is a 'Niṣkāmakarmayogi' or a 'sthitaprajñā' as described in Gitā.

To speak of means or the path of living liberation. Śaṅkara holds that it is knowledge alone, the real knowledge of the self, can lead to liberation. As bondage is caused by nescience i.e. defective knowledge of the self, when such defective knowledge is removed by real knowledge of the self, one attains liberation. According to him highest knowledge itself is liberation. All other means are subsidiary means. Devotion helps a person for his cittaśuddhi. karma is the remote cause of the rise of the desire to acquire the self knowledge. Aṣṭ āṅgayoga, he believes as internal means for the attainment of Jīvanmukti. Śravaṇa. manana and nididhyāsan also help a mumukṣu to procede towards his goal.

Maṅdana holds that the direct realization of the absolute unrelated Brahman i.e. Nirguna-Brahman-Sakṣātkara results from constant contemplation on the knowledge of the Truth received from the Upaniṣadic texts. Meditation on 'AUM', Sanyāsa all others are subsidiary means. For Sureśvara, knowledge is the sole means to attain Jīvanmukti. Vācaspati Mīsra also is of the same opinion that knowedge is the sole means and all other such as performance of prescribed duties sacrifices, charity, penance is preparatory to the emergence of knowledge. It produces the desire to know Brahman.

Thus nearly all the Advaitins hold jñāna-karma-sacrifices-yoga-śravaṇa-manana-nididyāsana-moral discipline-charity-penance etc. as the means to Jīvanmukti. Vidyāraṇya alone holds that yogic practices which is helpful for the Vāsanakṣaya and manonāśa is the whole and sole means to Jīvanmukti.

Knowledge for him is the subsidiary means. In fact all these means are equally important. All these are the different steps which take a person to the ultimate Supreme goal. Each and every step is important.

One may arrange them according to his own way. It may be Karma-jñāna-Bhakti-Tapas or the contrary to it, it makes no difference. One thing is certain which Vidyāraṇya states at the beginning of his work Jīvanmuktiviveka,' that One must renounce the world, the very day one feels complete detachment. I think the journey towards Jīvanmukti begins from this point. A person who continuously runs after the material things to gain pleasure, at some particular point he comes to known that nothing can make him fully satisfied A person who loves and sacrifices for his relatives,friends and expects from them as well same love and sacrifice, at certain point, he comes to know that all his expectations were in vain. Thus at certain stage of life, one may feel detachment in his mind of this world. Vidyāraṇya advices that as soon as one feels complete detachment one must renounce the world, the very day. Such a person goes to his Guru and asks. "Is there not a greater mission in life than eating and drinking, growing and dyeing, earning and spending ? Is life only a continuous story of disappointments ? Is there greater world of ampler possibilities ? Is there any noble field of unbroken peace and joy ? and the master asks him to sit near and unwinds gradually all the secrets of these questions and ultimately the disciple reaches at the stage where he knows the falsity of the pluralistic world and eternal truth of the Supreme Self He experiences the Brahman. The disciple experiences within himself that Truth which is non-dual and beyond all phenomena. Brahmānubhava is experienced only by such a thirsty aspirant. An aspirant has to take great pain and persue his goal with a great devotion, Once his goal is achieved i.e.once he becomes a Jīvanmukta, all his expectations disappear. He realises the truth that - expectations are never fulfilled. He becomes a living God on earth. He lives only for the betterment of others, not for himself each and every common man can not go to āśrama and ask his guru the path of 'Jīvanmukti'. but the work 'Jīvanmuktiviveka' in his hand performs the role of a guru and solves all the questions in his mind It plays the role of a lighthouse in the darkness of miseries and sorrows. It leads an aspriant to right direction. Thus Vidyāraṇya's contribution to Advaitic concept of living liberation proves to be valuable one. He proves importance of the concept with the help of Yogavāsiṭha, the philosophical work which deals with Vairāgya. mumukṣu-vyavahāra. Upāsanā, nirvāṇa etc. He also quotes verses from Gitā which guides one, how to reach the Brahman-state .Vidyāraṇya also quotes the precious verses and mahāvākyas from different Upaniṣads. Thus though the path of living liberation is called the toughest path to follow, Vidyāraṇya makes it easier for an aspirant.

Vidyāraṇya does not differ very much from Śaṅkara regarding the views on Jīvanmukti. He is more or less the same in treating the subject. Though Jīvanmukti is discussed in various Upaniṣads and Bhagavadgitā, it was Śaṅkara who advocated the concept clearly, emphasiszd the importance of Jīvanmukti and Videhamukti seperately. All other Advaitins accepted the concept but Vidyāraṇya systematized it. He gave a systematic account of Jīvanmukti. He has added many new details in the discussion of the problem.

Both Śaṅkara and Vidyāraṇya agree on the point that the aspirant should strive to attain the state of Jīvanmukti. In Vivekacūdāmaṇi Śaṅkara says. "A firm conviction of the mind to the effect that Brahman is real and the universe unreal, is designated as discrimination (viveka) between the real and unreal (20). There is no hope of immortality by means of riches - such indeed is the declaration of Vedas. Hence it is clear that works can not be the cause of Liberation (7). Therefore the man of learning should strive his best for liberation, having renounced his desire for pleasures from external objects, duty approaching a good and generous preceptor, and fixing his mind on the truth inculcated by him.

Śaṅkara lays more stress on the ethical aspect of the life of a Jīvanmukta in his Ātmabodha. In Vivekacudāmaṇi, he has explained how a Jīvanmukta should involve himself in the works of Lokasaṅgraha. His own life itself shows how, after attaining the highest spiritual illumination, he engaged himself in the active service of humanity going about the length and breadth of the vast Indian sub-continents in those days when travelling conditions were extremely difficult. But Vidyāraṇya thinks that even such a Lokasangraha, we may create bondage.

Śaṅkara and Vidyāraṇya differ regarding the exact connotation of videhamukti. Videhamukti according to Śaṅkara is meant as mukti which is attained after the decay of the body, but for the latter, it means the absence of future body (bhāvideha). Thus Vidyāraṇya identifies, in a way, Videhamukti with Jīvanmukti, because in both these muktis there is an absence of future body. It means the gurantee that there would be no future birth for a Jīvanmukta. Śaṅkara on the other hand, in his commentaries on Brahmasūtra accepts that Jīvanmuktas may reincarnate themselves as human beings with a view to help the people in their spiritual journey.[1]

Both Śaṅkara and Vidyāraṇya accept that Prārabdha Karma remains in name for a Jīvanmukta and that knowledge can not destroy these karmas but Vidyāraṇya goes a step further and states that the prārabdha, which remains in the name and does not conflict with one's self-knowledge, may be overcome by the practice of yoga in this very life.

Vidyaraṇya lays more stress on manonāśa and vaśanākṣaya. Śaṅkara also has added many verses in his Vivekacūḍāmaṇi that a Jīvanmukta must control his mind. He says : Attaining purity through a preponderance of discrimination (between self and not-self) and renunciation (of the not-self), the mind makes for liberation. Hence the wise seeker after Liberation must first strengthen these two (175).

1. Advaitic concept of J.M.: L.K.L. Shrivastava, p.277

In the forest-tract of sense-pleasures there prowls huge liger called the mind. Let the good people who have a longing for Liberaion never go there. (176)

But ultimately Both Saṅkara and Vidyāraṇya emphasize on the knowledge as the ultimate truth. When the aspirant realises that - There is neither death nor birth, neither bound nor a struggling soul, neither a seeker after Liberation nor a liberated one - this is the ultimate truth, (575) an aspirant attains mokṣa.

Thus all the Advaitins accept Jīvanmukti. Not only Advaitins but in Buddhism Arhat is Jīvanmukta. He is one who has overcome all sufferings and miseries of life. He has attained the knowledge of truth and is unattached with sense - pleasure and the evils of the ceaseless ocean of life and death. He is also called 'vitaragi in some Buddhist Texts. An Arhat is Sarvajña. He is mukta. He lives in the world only due to compassion for other people the bound and to experience past karmas. Thus Buddhism accepts Jīvanmukti.

Jīvanmukti is well depicted in Jainism also. Tirthankaras, who are worshipped by the Jains are Kevali.

They have bodies even after attaining Kevalajñāna or omniscience. It is with the help of this bodily existence that they are able to preach the dharma (Truth) to the people, because their main function is dharma-pravacana.

Thus Jīvanmukti is important as it bridges the life on earth with that which is beyond our perception. It shows the path how to live peacefully. He realises that there is neither death nor birth, neither a bound nor a struggling soul, neither a seeker after liberation nor a liberated one-this is the ultimate Truth.

Mokṣa in Advaita is the eternal communian with Brahman. Brahman which is unmanifest, indeterminate, imperishable, attributeless, self-luminious consciousness. It is a Supreme Reality. Mokṣa is merging of an individual self in such a Brahman. Mokṣa is eternal peace, freedom from all kinds of sorrow. It is an eternal bliss. The ultimate goal of human b ings —Mokṣa .

BIBLIOGRAPHY

Advaitic concept of Jīvanmukti by Shri L.K.L. Shrivastava, Bharatiya Vidyā Prakashan, Delhi, 1990

Advaita of Śankara : A Reappraisal, Dr. S. G. Mudgal, Motilal Banārasidas-Delhi, 1975

Ātman and Mokṣa, Dr. G. N. Joshi, Gujarat University, Ahmedabad, 1965

Anubhūtiprakash of Vidyāraṇya, Godabarish Mishra, University of Madras, 1992

Ācarya Śankara, Swami Apurvananda, (Marathi), Ramkrisna Math Nagpur, Tr. By Shri P. G. Sahasrabuddhe, 1995

Akalanka's Criticism of Dharmakirti's Philosophy, A study by Nagin Shah, (L D. Series II), Ahmedabad 1967

Brhadāraṇyaka Upanisad, Upanisad Series, Sixth Impression, Shri Ramakrisna Math, Mylapore

Brahmasūtra Śānkarabhāṣya, Vol. I, II, III, Dr. S.D. Deshmukh, (Marathi), Prasad Prakashan. Pune. Sake 1916

Bhāvārtha-Dipika, Jñanesvari By Shri Jñanadeva Published for Samta Books by V. Sadananda Madras

Bhaṛtrharinitiśatakam, The Century of life By Shri Aurobindo, Shri Aurobindo Institute of Research In Social Science Shri Aurobindo Society, Pondecherry, 1998

Brahmavāda, The System of Shuddhādvaita of Shri Vallabhācarya By Shri Raghunath Sharma, 1992 Published by Raghunath Goswāmī

Comparative Study of the concept of Liberation in Indian Philosophy, A. K. Lad Girdharlal Keshwadas Chawk, Burhanpur (M.P.), 1967

Constructive Survey of the Upanisadic Philosophy, R. D. Ranade Bharatiya Vidyā Bhavan, Bombay, 1968

Concept of Mukti in Advaita Vedānta, A. G. Krisna Warrier University of Madra, 1981, Series No. 9

Concept of Vyāvahārika in Advaita Vendānta, T. P. Ramacandran, University of Madras, 1980, Series No. 12

Die weg zur Erlosung bei Lebseiten Ihr Wesen and 1 rh Wert Nach den Jīvanmukti des Vidyāraṇya, Wiener Zeitschrift fur die kunde sud un Asiens 8 (1964) 224-62 and 17 (1970) 131-159, J. F. Sprockhoff.

Drg.-Ḍrsya-Viveka : Swami Nikhilananda, Shri Ramkrishna Ashram, Mysore 2, 6th ed., 1976

Foundation of Hindusim, Dr. Y. S. Shastri, Yogeshwar Prakasan, Ahemedabad, 1993

Gaudapāda : A Study in Early Vedānta, T.M.P. Mahadevan University of Madras, Madras, 1960

Glory of the Vedas, Dr. Sudyumna Acarya, Veda Vani Vitanam, Research Institute of Indology, Kolgavan, Satna (M.P.)

Gems from the Upanisads, H. B. Philips, Shri Ram Krisna Math, My'apore, Madras, 1977

History of Indian Philosophy, Vol. I & II, Jadunath Sinha, Central Book Agency, Calcutta, 1952

History of Indian Philosophy, Vol, I, II, III, S.N. Dasgupta, Motilal Banarasidas, 1975

Hymns of Śaṅkara, T.M.P. Mahadevan, Motilal Banarasidas Publishers Pvt. Ltd., Delhi, 1980

Indian Philosophy, Vol. I & II, Dr. S. Radhakrisnan, Centenary Edition, Oxford University Press, Bombay, Sixth Impression, 1993

Introduction to Indian Philosophy, S. Chatterjee and D. M. Datta, University of Calculla, 1960

Īśadi-Astottarasatopanisad, Ed. Vasudev Sharma Panasikar Nirṇaysagar Press Bombay 1917

Jīvanmukti in Transformation, Embodied Liberation in Advaita and Neo Vedānta, Andrew O Fort, Albany State University of New York Press, 1998

Jīvanmuktiviveka (Liberation in Life) of Shri Vidyāraṇyaswāmī, Edited with English Translation by Pandit S. Subramnya Shashtri and T. R. Shrinivas Ayangar, The Adyar Library and Research Centre, Madra, 1978

Kashmir Saivism, L. N. Sharma, Bharatiya Vidya Prakasan, 2nd Ed., Delhi, 1996

Living Liberation in Hindu Thought, Andrew O Fort and Patricia Y. Hume, Albany State University of New York Press, 1996

Liberation while living in Jīvanmuktiviveka, A chapter on Vidyāraṇya's Yogic Advaita, Andrew O Fort, Texas Christian University

Lectures on the Ramayana, The Rt. Hon. V. S. Shrinivasa Shastri, Madras, Sanskrit Academy by S. Vishwanathan, 1977

Mahayanasutralankara of Asanga, Dr. Y. S. Shastri, Shri Sadguru Prakasan, Delhi, 1989

Māṇḍukya Upaniṣad, Gaudapadakarika, Swami Chinmayananda, Central Chinmay Mission Trust, Bombay, 1990

Naradasmriti : A Study by Dr. Sunanda Y. Shastri, Ph. D. Thesis Gujarat University, 1991

Outlines of Indian Philosophy, Jadunath Sinha, Sinhna Publishing House Pvt. Ltd., Calcutta, 1963

Pancada' si, com. By Swami Suddhabodhānanda, Shri Viśverwar Trust, Bombay

Philosophy of Advaita, T.M.P. Mahadevan, Arnold, Heinemann, 1976

The Principal Upaniṣads, Dr. S. Radhakrisnan, Harper Collins Publishers, India, 1994, Daryaganj, New Delhi, Eighth Impression, 2000

Primal Spirituality of the Vedas : Its Renewal and Renaissance, R. Balsubramania, 1996, Published by Prof. Bhuvan Chandel, Project Co-ordinator for PHISPC, Delhi-92

Sankara's Brahmavāda, Dr. R. S. Naulakha, Kitab Ghar, Ācarya Nagar, Kanpur, 1964

Sāmkhya-Kārika, Iswarkrisna Com.by Gaṇḍapāda Eng. Tr. By Colebrooke

Salient Features of Hinduism, Dr. Yajneswar S. Shastri, Yogeshwar Prakashan, Ahmedabad, 1994

Shri Madbhagawadgita, A.C. Bhaktivedanta, Swami Prabhupada, Shrila Prabhupada's Centenary, Marathi Tr. By J. G. Karandikar, 1991

Sārth Jñaneswari : Jñānadev, Balkrisna Ananta Bhide, April 1991

Sanskrit for Beginners, Dr. Sunanda Y. Shastri, Yoeshwar Prakashan, Ahmedabad, 1994

Śankara, Digvijaya : Vidyāraṇyaswāmī, Gujarati tr. by Shri Krisnalal Govindaram Deva'sray, Purusottam Kahnaji Gandhi, Shardapith Vidyā Sabha, Dwarka, 1979

Shri Brahmasamihita : His Divine Grace Bhaktisiddhanta, Saraswati Goswāmī Thakur, 2001, 1972, The Bhaktivedanta Book Trust International INC.

Śankara-Digvijaya : Madhava-Vidyāraṇya Tr. By Swami Tapasyananda, Shri Ramkrisna Math, Madras, India

Tarkasangraha, Annambhatt, Gujarat tr. Tr. By Jitendra S. Jetley, Ahmedabad, 1963

Ten Principle Upaniṣads with Saṅkarabhāṣya By Motilal Banarasidas, Delhi, 2nd Ed., 1978

Teaching of Upaniṣads, Dr. Sunanda Shastri Sharda, Jnanapeetham, Pune, First Ed., 2004

Upanisads Retold, V. H. Date, Vol. I & II, Minishram Manoharlal Publisher Pvt. Ltd., 1999

Upaniṣads Translated by Swami Nikhitananda, Phoenix House Ltd., LONDON

Vivekacudamani (Tr.) Mohini M. Chattergi, Theosophical Publishing House, Adyar, Madras, India, 1983

Vedānta Philosophy, Max Muller, Cosmo Publications, 24-B, Ansari Road, Daryaganj, New Delhi, 1965

Vedānta : A Culmination of Thought, R.D. Ranade, Bharatiya Vidyabhavan, Bombay, 1970

The Vedānta, V.S. Ghate, V.G. Paranjpe, C.R. Deodhar, Bhandarkar Orient Research Institute, 1960

GLOSSARY

Abhaya— fearless

Acala— steadfast, immovable

Achintya— unthinkable

Adharma— demerits

Adṛaṣta— karma

Āgamas— kanuals of worship, they are also called 'Tantras'.

Ahiṁsa— non-violence

Ajiva— matter, jada (thing enjoyed)

Ālaya Vijñāna— (Pure Being, or Store Consciousness)

This school has been ientified with the absolute citta-stuff and made the foundation of the whole Vijñāna Group. This constitutes the distinctive concept of te Vijñāna-vāda School.

Anantacatuṣtasya—Anant means infinit

The soul in its pure state (i.e in the state of Mokṣ) is of infinit perception (ananta darśana) infinit knowladge (ananta jñāna) infinit bliss (ananta sukha) infinit power (ananta vīrya) so it is all perfect.

Ānanda— joy

Anekānta— The theory of infiniteness (jivas are infinit in number)

Anivṛttikarana Guṇasthāna— This is the nineth Guṇasththāna in the ladder of fourteen steps towa-rds Mokṣa in Jainism. At this stage the soul performs the advance upto the point of spiritual Vision.

Antaḥkaraṇa— This type of action is responsible for causing the puzzle and unwillingness of the mind of person to let go important and favourable opportunities for their uplift and betterment.

Anubhava— experience

Anubhāga bandha—Bondage (bandha) depends upon intensity (anubhāga) It may be fast or loose, sharp or acute with angularities.

Anubhūti— Spiritual experience

Apramatta Samyata Guṇasthāna—This is the seventh stage of the ladder of fourteen steps of Guṇa-sthānas in Jainism. In this stage the spiritual inertia of the asprirant is also overcome. It is the stage of self-control with freedom.

Apurvakaraṇa or Nivrtti Guṇasthāna—This is the eight stage of the ladder of fourteen steps of Guṇa-shānas in Jainism. At this stage soul attains special purification and finds itself competent to a\reduce thẹ duration, and intensity of previously acquired Karmas. Here an aspirant increasses his power of meditation by yoga and fetters of Karmas are loosened fast.

Arhat— A perfect person. One who reaches at the Nirvāṇa stage is called Arhat.

Asamprajñāta Samadhi—Unconscious variety of trance.

Āsana— Posture.

Ascetic Yogin— A realized renunciator.

Āsrava— Karma current, that enter the soul and cover it, which becomes the cause of bondage.

Asura— Demons.

Āstika— One who believes in God.

Āvaraṇaśakti— Concealing power.

Avidyā— Nescience.

Avirata Samyag Driṣti Guṇasthāna—This is the fouth stage in the ladder of fourteen Guṇasthānas in Jainism. This is the stage which is right vision without abstinence.

Ayogalevali Guṇasthāna—This is the last, the fourteenth stage in the ladder of Guṇasthānas in Jaini-sm. This state is absolute motionless. At this stage soul attains disembodied emanciption or videhamukti.

Āyu Karma— This kind of action fixes the length of time for a jiva to spend in a particular form which has benn given to it, by its Karma (action) it determines the span of life of the Gods, human beings and the animals of the lower species.

Bahudaka— Renuncient having sharper detachment.

Bhāvamokṣa— Freedom from physical bondage, Jivan-mukti, living liberation.

Bhāvabandha— The physical condition which causes bon-dage is called Bhāvabandha. It arises from false belief, want of character etc.

Bhakta— A devotee.

Brahman— Supreme Self.

Brāhmana— Knower of Supreme Self.

Brahmaloka— Loka means sphere or world. Brahman is Supreme Self. So it means sphere of Sup-reme Self.

Brahmachāri— A student.

Cāndāla— Unholy Person.

Cit— Intelligence.

Citta— Mind.

Darśanāvaranīya—Intution obscuring Karmas.

Dehatmavādins— Believers in body as soul.

Desavrjāti Guṇasthāna—This is the fifth stage in the lad-der of fourteen Guṇasthānas in Jainism. At this stage soul achieves right vision And capisity for partial abstinence.

Dharmakāya— Divine Buddha.

Dhrma— Religious merits, Moral virtues.

Dhārana— Absorption of mind.

Dravya— Elements.

Dravyamokṣa— Actual separation of karmas from the soul *i.e.* Videhamukti (liberation after the fall of body.).

Dukhkṣaya— End of miseries, Destruction of miseries.

Ekarasa— Same nature.

Gotrakarma— This kind of action (karma) determines the cast in which a jive may be born and spend its life throughout the period of its existence.

Grahastha— A householder.

Granthibheda or Saswadana Samyagdrsti—Guṇasthān is the second steps in ladder of fourteen Guṇasthāna in Jainism. Here the soul halts during its fall a higher stage.

Guṇasthāna— Spiritual Development.

Jiva or Jivātman—Individual Self.

Jñānaswarupa— In the form of knowledge.

Jñānāvaraṇīya— Intution obscuring actions.

Karma— Action.

Karmabeeja— Merits and demerits.

Kaṣāya— Passion.

Kṣīṇakaṣāya Guṇasthāna—This is the twelveth stage of the ladder of fourteen Guṇasthānas in Jainism. At this stage passions are annihilated. The soul is free from all types of obscuring karmas.

Kuticaka— Renuncient having sharp detachment.

Lokeṣaṇā— Desire for the world.

Matha— Monestry.

Manna— Reflection.

Misra Guṇasthāna or Samyag Mithyādrsti Guṇasthāna—This is the third Guṇasthāna in the ladder of fourteen guṇasthānas in Jainism. At this stage soul combines both right and wrong attitude.

Mithyādṛṣti Guṇasthāna—This is the first stage of the ladd-er of fourteen Guṇasthāna and the lowest stage of perversity of attitude towards truth. Here the soul has minimum possible degree of right vision.

Mithyātva— Subreption.

Mithyājñāna— Misapprehension.

Mohanīya karma—This type of action (karma) causes in us infatuation by means of worldy attachments and indulgence of our passions.

Mokṣa— Liberation.

Mukta jivas— Liberated souls.

Mumukṣu— A seeker of liberation.

Nāstika— One who disbelieves in God.

Nididhyāsana— Assimilation.

Nirguṇa Brahman—Attrubuteless Higher Brahman.

Nirvikalpa Samādhi—Absolute interception.

Niscalāsthiti— State of equanimity.

Niyama— Observance.

Parāvairagya— Supreme detachment.

Pariṇāma— Transformation.

Pradesabandha— This refers to the different parts of the soul to which the karma particles. (bandh means bondage)

Prakrtibandha— Bondage acording to one's nature.

Pramatta Samyatā Guṇasthāna—This is the sixth stage of the ladder of Fourteen Guṇasthānas in Jainism. At this stage soul remains spiritual inertia (pramāda).

Prāṇāyāma— Restraint of breath.

Pudgala— Matter.

Rajas— (Rajoguṇa) Constituent aspect of mind, compact of all passions and sourse of longing and attachment. It fetters down the body owner by fondness for actions. Thus from rajas greed is born.

Saguṇa Bhakti— Idol worship.

Saguṇa Brahma—God, Īśvara, Possessed of auspicious qualities.

Sādhaka— Aspirant.

Sāksaātkāra— Intution.

Śakti— Power.

Samādhi— Concentration.

Saṁsāra— Phenominal world.

Saṁvara— To check karma current is Saṁvara (control of karma current).

Samsāri Jīva— Worldly souls, Souis in bondage.

Samprajñāta Samadhi—Concentration of the conscious variety.

Sāmya— Sameness, State of equality.

Sancita karma— Accumulated actions.

Sanyāsi— A renuncient.

Sarvajñatā— Omniscience.

Sat— Existence.

Sattva— Constituent aspect of mind which is auspicious, taintless, huminious and free from disorder. From sattva wisdom is born.

Sayogakevali Guṇsthāna—This is the thirteenth stage of the ladder of fourteen Guṇasthānas which is in Jainism equivalent to jivanmukti state of other schools.

Sāyujyatā— Penetration, Uninion without duality.

Siddha— A perfect being.

Siddha jīva— Liberated soul.

Śravaṇa— Listening (holy literature).

Sthitaprajña— One whose intellect, wisdom is stable.

Sthitibandha— Bondage (bandha) according to duration. It may last for any length of time, extending even to thousands of years.

Sukṣma Samparāya Guṇasthāna—This is the tenth stage of the ladder of fourteen Guṇasthānas . At this stage only the subtle greed may be the form of subconscious attachment to the body.

Swarupa Lakṣana—Indeterminate, Attributeless.

Syādavāda— It means reality is indefinite. The point stressed in Syādavāda is that all knowledge is probable (Syad) i.e may be.

Tamas (Tamoguna)—Constituent aspect of mind which is born of nescience, which causes infatuation to all body-owner. Through heedlessness, indolence and slumber, it fetters down. Thus from tamas delusion is born.

Tatastha Lakṣaṇa—Accidental qualities.

Tirthandara— Profets of Jainism.

Tri-ratna— Three jewels of Jainism, Right faith, Right knowledge and Right conduct, these three prepare the path to liberation.

Upādhi— Delimiting adjuncts.

Upāsanā— Path of devotion.

Uttamasukh— Highest bliss.

Uṣatamoha or Upasantakaṣāya Guṇasthāna—This is the eleventh stage of the ladder of fourteen guṇasthānas. At this stage even the subtle greed is suppressed and the soul is free from passion.

Vānaprastha— Resident in hermitage.

Varṇāśama dharma—Varna (cast) āśrama (stage). Duties belongs to cast and stage.

Vedanity karma—This kind of action makes us experi-ence pleasant and sweet fruits or bitter and painful results of our previous deeds.

Veetarāga— Without attachment.

Veetarāgi— One who holds pleasure and pain alike.

Veetaśoka— Free from grief.

Vidyā— Intuition.

Vivarta— Appearance.

Vritti— Mental Manifestation.

Vyāvaharika Sattā—Phenominal Plane.

Vyavahrika Point—Empirical point of view.

Yama— Forbearance.

Yoga— Meditation of God.

INDEX

A

Absolute-12, 106, 117, 119, 145, 164, 165, 182, 190
Absolute Idealism-105
Absolute Reality-107
Absolutism-77
Absolutist-29
Accidentalists-23
Acintya-bhedabhedavada-74, 78
Adharma-35, 58, 174
Adjunct-78
Adrsta-57, 99, 209
Advaita Vedanta-11, 13, 15, 18, 47, 66, 78, 97, 104, 105, 114, 116, 120, 126, 128, 134, 179
Agamapramany-64, 77
Agamas-76, 77, 78, 79, 81
Agamashastra-104
Aham Brahmasmi-149
Ahamkara-48
Ahimsa-34,
Ajitanatha-34
Ajiva-35, 36
Alatacakra-97
Alatasanti-97
Alaya-30
Amrtatva-85
Ananda-71, 107
Anandagiri-126
Anandamaya-98
Anandanubhava-125
Anantacatustaya-41
Anekanta-34
Anirvacaniya-123
Anubhutiprakasa-132, 136
Anupadhisesa-
Aparoksanubhutitika-132
Apavarga-49, 53, 54, 58
Appaya Diksita- 12
Apratisthita Nirvana-32
Aranyakas-93, 149
Arhat-31, 32, 215
Arjuna-87, 88
Art of Living- 92
Artha-1,
Asanga-30
Asparsa Yoga-103
Asrama-14, 157, 192
Asrava-37
Astanga Yoga-55, 185, 211
Ativarnasrami-14, 141, 144, 157
Atman-1, 2, 7, 8, 13, 24, 26, 85, 86, 95, 96, 97, 99, 101, 102, 106, 107, 110, 111, 116, 156, 163, 167, 168, 173, 176
Atma-saksatkara-1
Atharva Veda-4
Atheism-51

Athestic-51
Atheists-23, 25
Atmabodh-214
Aum-98, 99, 100, 101, 120
Aurobindo-87
Avidya- 4, 66, 70, 73, 108, 112, 113, 119, 121, 122, 210, 113, 122, 123, 124, 126, 136, 137, 141, 168, 169, 210

B

Badarayana- 93, 96,
Baddha Jivas-65
Baladeva-75
Basavesvara-81
Bhagavad Gita-14, 21, 64, 67, 71, 87, 103, 143, 153, 155, 156, 165, 173, 174, 175, 184, 186, 190, 196, 201, 204, 210, 213,
Bhagavatapurana-71, 76
Bhakti-71, 73, 74, 76, 78, 83, 90, 116
Bhaktiyoga-66, 89, 90
Bhamati-121, 122, 125
Bharatitirtha-130, 131
Bhaskara-62, 63, 78, 93, 128
Bhedabhedavada-62, 78, 128
Bhiksuki-146
Bhiksus-28
Birth-2, 6, 10, 26, 38, 48, 52, 109, 173, 209, 215
Bliss-3, 4, 21, 67, 70, 71, 73, 74, 83, 85, 99, 104, 135, 140, 157, 186, 192, 193, 197, 201, 202, 207, 208, 215
Bodhisattva-31, 33
Bondage-2, 14, 15, 23, 28, 36, 37, 38, 44, 47, 50, 52, 53, 63, 66, 70, 73, 76, 78, 83, 108, 118, 125, 135, 150, 164, 165, 166, 167, 179, 180, 181, 211,
Brahma-78, 95, 199, 204
Brahmajnana-113, 114
Brahmaloka-94, 95, 145, 151
Brahman-2, 7, 8, 9, 12, 13, 14, 15, 16, 17, 19, 30, 62, 63, 65, 67, 70, 71, 72, 73, 75, 82, 83, 84, 85, 89, 90, 92, 93, 94, 95, 96, 97, 99, 101, 104, 106, 107, 108, 110, 111, 112, 115, 116, 117, 123, 124, 132, 136, 137, 138, 139, 140, 141, 142, 151, 153, 154, 155, 156, 163, 165, 168, 169, 174, 179, 180, 193, 197, 199, 201, 202, 204, 207, 209, 210, 211, 212, 213
Brahmananda-110
Brahmanas-79, 156
Brahmanubhava-2, 8, 109, 110
Brahmanubhuti-110
Brahmaparinamavada-62
Brahmasaksatkara-
Brahmasiddhi-121, 122
Brahmasutra-62, 64, 67, 71, 92, 93 94, 96, 106, 214

Brahmavidya-93, 131, 192, 197
Brahmopanisad-134
Brhadaranyaka Upanisad-134, 136
Brhaspati-22
Buddha-26, 28, 30, 32
Buddhahood-30
Buddhatva-33
Buddhism-22, 25, 27, 28, 33, 67, 78, 98, 105, 215
Buddhist-32, 46, 144
Buddhistic-
Philosophy-25, 27, 51
Bukka I-130, 131

C

Caitanya-74, 75, 78, 93
Caitanycaritamrta-74
Caturdharitika of Moksadharma- 146
Carvaka-22, 23, 24
Chandogya Upanisad-9, 119, 134, 136, 199
Cit-71, 107
Citta-30
Chitsukhacharya-125, 126, 210
Citsukhi-126
Cognition-44, 56, 102, 140
Concentration-192, 198
Consciousness-26, 30, 35, 36, 45, 62, 66, 83, 86, 99, 102, 108, 136 163, 168, 171, 185, 201, 204, 207, 208
Creator-75, 95
Creation-67, 95

D

Dark Night of the soul-88
Dasamula Sloka-74
Dasyarasa-
Death-2, 6, 7, 10, 14, 23, 25, 32, 38, 48, 52, 109, 157, 170, 174, 209, 215
Demerits- 58, 122, 123,
Devayana-13
Dharma-1, 35, 58, 174, 198, 215
Dharmadhatu-30
Dharmakaya-30
Dharmanairatmya-29
Dharana-189
Dhurtacarvaka-23
Dhyana-189
Difference and Non-Difference-70
Divine-74, 75, 82, 85, 108, 167
Drk-Drsya-Viveka-132, 141
Dualism-51, 52, 67, 72
Dualistic Non-
Dualism
Duality-30, 97, 101, 108, 124, 209
Dvaitadvaita-78, 128
Dvaita Vedanta-67, 78, 128

E

Efficient Cause-64
Egoism-48, 51, 155, 157, 172

Emancipation-1, 3, 80
Enlightened-25, 170
Enlightenment-32
Epistemology-42
Eternal Bliss-86
Eternal Entity-56

F

Final Beatitude-
Finite-66
Freedom-

G

Ganganatha Jha- 51
Gaudapada-97, 98, 101, 103, 104, 105, 106, 210
Gauranga-74
Gautama Buddha-25
Gitartha Samgraha-64
God-7, 14, 15, 22, 25, 34, 51, 55, 64, 65, 71, 73, 74, 75, 76, 77, 78, 81, 84, 85, 88, 89, 90, 94, 103, 104, 113, 115, 118, 156, 157, 198, 199, 201, 209, 213
God Realization-101
Gopala Krsna-70
Goswami-74
Gotama-42, 48
Govindabhasya-74
Govindapada-98, 105, 106
Govindayati-97
Gradual Release-92
Gunatita-14, 126, 156

H

Harihara I- 130
Hastamalaka-
Heaven-23, 84, 139, 158
Hedonists-23
Hell-23, 158
Hindu-42, 87, 131, 134
Hinduism-1, 2, 76, 87, 105
Hinayana-26, 27
Hinayanism-26
Hinayanist-27, 29, 32
Hiranyagarbha-13, 201

I

Ignorance-2, 4, 8, 28, 31, 38, 47, 53, 57, 63, 73, 80, 81, 83, 90, 108, 110, 111, 113, 119, 138, 140, 158, 168, 170, 171, 172, 173, 185, 192, 196
Immortality-6, 9, 25, 171, 174, 209, 213
Incarnation-173, 174, 186, 194, 198, 200
Independent Reality-
Indescribable-30
Indian Philosophy-1, 2, 10, 25, 33, 36, 42, 52, 53
Individual Self-9, 46, 63, 98, 102, 104, 108
Individual Souls-70, 83, 108
Infinite-66
Infinite Substratum-65
Innumerable Souls-
Intellect-100
Isvarakrsna-54

Isvara-12, 16, 95, 96, 101, 158

J

Jaimini-55, 58
Jaimini-Nyaya-Mala-Vistara-129
Jaina-41
Jainism-11, 22, 34, 37, 39, 40, 78, 215
Janak-179
Jina-34
Jīva-12, 16, 35, 36, 70, 72, 73, 84, 78, 81, 96, 124, 152, 210, 211
Jīvanmuktiviveka-20, 127, 132, 133, 143, 144, 195, 203, 212, 213
Jīvanmukta-14, 42, 55, 82, 104, 112, 113, 118, 124, 127, 131, 137, 141, 144, 153, 154, 170, 180, 181, 208, 210, 211, 212
Jīvanmukti-11, 12, 13, 14, 20, 21, 31, 33, 50, 54, 61, 63, 83, 104, 113, 114, 115, 118, 119, 125, 127, 128, 135, 136, 143, 144, 150, 151, 153159, 160, 161, 165, 166, 168, 181, 182, 193, 194, 195, 197, 198, 200, 201, 209, 210, 211, 212
Jivatman-46
Jnanakanda-55
Jnanakarmasamuccaya-63
Jnanasvarupa-70
Jnanavadin-
Jnanayoga-66, 89
Jneyavarana-29, 30

K

Kaivalya-53, 54, 168, 171, 200
Kaivalyopanisad-146
Kala-72
Kama-1,
Kanada-42, 44
Kapil-51, 52
Karma-2, 3, 23, 25, 35, 37, 39, 58, 60, 65, 66, 70, 71, 78, 80, 109, 113, 118, 119, 120, 123, 124, 125, 158, 168, 169, 171, 209, 211, 212, 215
Karmabija-109
Karmakanda-55
Karma (Karmic) matter- 36, 37, 38, 48, 50, 57, 59
Karma-yoga-66, 70, 78, 89, 90, 115
Karmic-36
Karyabrahma-117
Kashmir Saivism-11, 80, 81
Katha Upanisad-134, 190, 194
Kevali-42, 215
Klesavarana-29, 30
Knowledge-7, 10, 11, 13, 15, 16, 18, 23, 28, 29, 36, 46, 47, 70, 71, 73, 74, 78, 79, 80, 84,85, 89, 90,

94, 95, 103, 104, 107, 108, 110, 112, 113, 115, 116, 117, 118, 119, 120, 121, 123, 124, 126, 136, 138, 140, 144, 145, 156, 163, 164, 165, 167, 168, 169, 170, 171, 171, 175, 176, 179, 181, 182, 184, 186, 193, 194, 195, 196, 197, 198, 199, 201, 202, 204, 208, 209, 210, 211, 212, 215
Kramamukti-11, 12, 13, 92, 116, 117
Krsnacaitanya-74
Krsna Cult-72
Krsnadas-74
Kumarilabhatta-55, 56, 57, 58, 60, 61
Kutasthanitya-109

L

Laghu-yoga-vasistha-20, 143, 210
Lankavatara-30
Laksmi-77
Liberation-1, 9, 10, 11, 12, 13, 15, 30, 37, 38, 39, 41, 44, 47, 49, 50, 53, 57, 59, 63, 70, 73, 75, 78, 81, 85, 92, 94, 95, 96, 98, 102, 104, 106, 108, 109, 112, 113, 115, 116, 117, 118, 121, 123, 124, 125, 135, 137, 144, 145, 150, 153, 158, 165, 166, 167, 168, 170, 180, 181, 184, 186, 203, 211, 215
Lingayata-81
Living Liberation-159, 161
Logic-42, 43, 44, 47,
Lokamanya Tilak-113
Lord Krsna-21, 73, 75, 87, 89, 90, 92, 155, 190

M

Madhura Bhakti-75
Madhavacarya-78, 129, 130, 131
Madhvacarya-67, 78, 93, 95, 128
Madhyamikakarika-31
Madhyamikas-27, 29
Mahabharata-79, 87, 156
Mahasakti, 83
Mahatma Gandhi-87
Mahavakyas-16
Mahavira-34
Mahat-190
Mahayana-26, 32
Mahayanist-27, 29, 33
Mahayanism-27
Mandana Misra-55, 118, 119, 120, 121, 122, 123, 210
Mandukya Upanisad-97, 98, 134
Manan-16, 17, 123, 125, 141, 146, 164
Manas-176
Manonasa-20, 160, 167, 168, 172, 194, 212

Manu-42
Manusmrti-164
Maryadamargiya-73
Material cause- 64, 94
Matter-35, 52
Mandukya Karika-98, 104
Materialist-23
Maya-72, 75, 80, 81, 83, 97, 101, 111, 122, 137, 158
Mayasakti-75
Mayavada-96
Meditation-29, 88, 120
Merit-38, 123
Metaphysics-42
Mimamsa-55, 56
Mimamsakas-57, 58, 61
Mind-30, 36, 100, 150, 151, 152, 155, 162, 166, 168, 176, 177, 178, 180, 181, 182, 183, 184, 185, 186, 187, 188, 190, 191, 193, 194, 197, 198, 208, 214
Mithyajnana-48
Mokṣa-1, 2, 3, 4, 6, 7, 10, 19, 25, 31, 33, 38, 49, 53, 62, 66, 71, 76, 78, 80, 81, 83, 86, 92, 109, 110, 111, 116, 128, 200, 209, 215
Mokṣasastra-2
Moksavastha-9
Mukta-58, 70, 73, 122
Mukti-2, 3, 13, 15, 76, 80, 85, 117, 121
Muktikopanisad-98
Mundaka Upanisad-7, 8, 85, 108, 134, 136
Mumuksu-18, 66, 118, 139, 161
Mumuksatva-179
Mundane Space-41
Murti, T. R.V.-31
Mysticism-82

N

Naiskarmyasiddhi-120, 122, 134, 163
Naiyayikas-46, 47
Nagarjuna-29
Narayan-64, 77, 78, 98
Naturalist-23
Nescience- 38, 66,112, 118, 119, 209, 211
Nididhyasana-16, 17, 123, 125, 141, 146, 164
Nihilist-27
Nimbarkacarya-70, 71, 78, 93, 95, 128
Nimai-74
Nirguna- 70, 96, 111, 211
Nirjara-39
Nirvana- 10, 27, 28, 29, 30, 31, 32, 39, 40, 213
Nirvikalpa-190
Niskama Karma- 90, 115
Nissreyasa-49
Nitya Jivas-65
Noumena-154
Non-Dualism-72
Non-Dualistic-30
Non-Dual spirit-64
Non-Duality-85, 98, 165
Nyaya-47

Nyayamalavistara-131
Nyaya Philosophy- 42
Nyayaratnakara-60
Nyaya-Vaisesika- 42, 43, 44, 46, 48, 50, 53, 59, 61, 128

O

Omniscience-42
Ontological Reality-107

P

Pandmapada-140
Pancadasi-131, 132, 139
Pancaratra Agamas- 77
Paramahamsa-21, 138, 145, 147, 148, 203, 204, 205, 206, 208
Paramahamsa Upanisad-134
Paramahamsa Yogin-14
Paramarthika Satya-95
Paramatman-46, 70, 71, 204
Parasara-98
Parasaramadhaviya-129
Parinamavada-65, 72
Parthasarathi Misra-55, 58
Parvati-82
Patanjali-54, 143, 162, 176, 185, 187
Phenomena-104, 107
Pluralism-52
Plurality-106, 111
Politheistic-56
Positivists-23
Prabhakara-55, 57, 58, 61
Prakriti-51, 52, 54, 70, 72, 93
Prana-185, 188
Pranava-85, 188
Pranayama-187, 188, 189
Prarabdha Karma-115
Prasthanatrayi-62, 84
Pratyabhijna-81
Pratyahara-189
Premabhakti-74
Primoradial Matter-52
Pudgala-36
Pudgalanairatmya-29
Purusa-51, 52, 54, 93
Purusottama-64, 72, 74
Purvamimamsa-22, 55, 56, 128
Pustimargiya-73, 74

Q

Qualified Monism-63

R

Radha-70
Radhakrishnan, S.-12, 23, 58, 111
Rajas-156, 175, 176
Rajayoga-89, 91, 152
Ramanuja-50, 63, 64, 65, 70, 72, 78, 93, 95, 128
Rama-78, 151
Rasalila-74
Realism-67
Realistic Pluralism-67
Reality-16, 26, 29, 30, 47, 70, 72, 78, 83, 97, 102, 103, 163, 197, 200, 210
Realization-1, 4, 63, 66, 73, 88, 89, 91, 96, 103, 104,

109, 110, 118, 119, 135, 136, 153, 155, 192
Rebirth-23, 40, 76, 109, 115, 186
Release-23, 47, 49, 66, 71, 74, 92, 119, 123, 141
Renunciation-125, 134, 136, 140, 144, 145, 148
Rgveda-4, 5, 130
Ritualism-24
Rsabhadeva-34
Rudra-79

S

Sabarasvami-55, 56, 58
Saccidananda-104
Sacrificialism-24
Saddarsanasamuccaya-22
Sadyomukti-11, 12, 13, 116, 117, 121
Saguna-70, 83, 96, 116
Saivas-62, 76
Saivasiddhanta-11, 79
Saivism-78, 79, 80
Saktas- 76, 77, 83
Sakti-81, 82, 83
Saktivisisthadvaita-81,
Salokya-10, 11, 12
Salvation-76
Samadhi-142, 190, 192, 193, 203
Samaveda-4
Samipya-11, 12
Samsara-29, 32, 49, 58, 70, 106, 108, 138, 183, 209
Samvara-37
Sankhya-50, 51, 144
Sankhyayoga-50, 52, 53
Sancit Karma-61, 115
Sanciyamana Karma-115
Sankara-12, 13, 18, 19, 64, 65, 70, 72, 84, 93, 95, 96, 97, 98, 105, 106, 107, 109, 110, 111, 112, 113, 115, 117, 119, 120, 121, 122, 126, 128, 136, 137, 138, 158, 184, 205, 210, 211, 213, 215
Sankarananda-130
Sankhya-31, 52, 54, 93
Sankhyakarika-54
Sankhyasutra-54
Sankhya Yoga-11, 50, 52, 53, 128
Sannysa-92, 120, 121, 127, 129, 130, 134, 147, 211
Sarvadarsanasangraha-22
Sarvastivadins-27
Satapatha Brahmana-5
Sarupya-10, 11, 12
Sat-71, 107
Sattva-156, 175, 176, 189
Sautrantikas-27
Savikalpa-190
Sayana-22, 130
Sayujya-10, 11, 12, 74
Science of Reality-92
Self-14, 15, 16, 43, 44, 53, 56, 89, 100, 118, 120, 172, 193, 202, 207
Self Realization-2, 10, 15, 102, 103, 120, 123, 145, 209

Sensualism-24
Shuddhadvaita-78
Siddhantarahasya-71
Siddhartha-25
Siddhitraya-64
Siva-77, 79, 80, 81, 82
Slokavarika-60
Smrit-124, 134, 144, 172, 175, 199, 204
Soul-1, 6, 7, 8, 14, 32, 35, 36, 37, 38, 40, 44, 45, 48, 53, 57, 63, 70, 71, 73, 74,75, 78, 80, 81, 82, 88, 91, 94, 136
Spirit-52, 150
Spiritual-128
Spiritual Experience
Spirityual Identity-138
Sravana-16, 17, 123, 125, 141, 146, 164
Sribhasya-64
Srikrsna-72
Srividya-82
Sruti-13, 16, 17, 43, 52, 87, 97, 105, 106, 110, 124, 134, 144, 166, 167, 169, 172, 175, 176, 188, 198, 202
Sthitaprajna-14, 113, 119, 126, 155, 156, 157, 211
Substance-45
Suddhadvaita-72, 128
Supramoral-111
Supramundane Spirit-36
Supreme Bliss-82, 136
Supreme Brahman-72
Supreme Knowledge-142
Supreme Person-72
Supreme Reality-75, 82, 215
Supreme Self-9, 19, 46, 66, 81, 85, 98, 100, 104, 108, 142, 205, 206, 208, 209, 212
Supreme Truth-102, 146
Suresvaracharya-118, 120
Svarupalaksana-174
Sydvada-34

T

Taittiriya Upanisad-7, 84, 136
Tamas-156, 175, 176
Tantrika Agamas-82
Tarka-42, 47
Tatasthalaksana-174
Tathata-30
Tattvamasi-137
Theism-77
Theistic-42, 72, 96
Theory of Causation-65
Tirodhanasakti-80
Tirthankara-38
Totakacarya-98
Transcendental Truth-107
Transmigration-109
Truth-4, 21, 99, 101, 103, 114, 120, 138, 155, 160, 164, 168, 211, 212, 215

U

Ultimate Reality-30, 70, 78, 128, 136,
Universal Self-91, 102

Upanisads-6, 24, 31, 51, 64, 67, 71, 83, 84, 87, 93, 105, 106, 108, 139, 149, 209
Uttara Mimamsa-55, 128

V

Vacaspatimisra-54, 210, 211
Vaibhasikas-27
Vaikuntha-65
Vairagya-179
Vaisnava-63, 76, 77, 78
Vaisnava Vedantins-50, 62, 111
Vaisnavism-64, 78, 79
Vaisvanara-99, 100
Vallabhacarya-71, 72, 73, 74, 78, 93, 128
Vardhamana-33, 34
Varna-14, 157, 192
Varnasramadharma-19, 116
Vasana-154
Vasanaksaya-20, 160, 161, 167, 168, 172, 181, 194, 212
Vasistha-98, 151, 158, 180, 183, 184, 185, 191, 192, 197, 200
Vasudeva-64, 77, 78
Vatsyayana-47, 50
Vedanta-1, 14, 16, 31, 38, 47, 61, 64, 66, 67, 72, 101, 128, 139, 142
Vedantapaksa-125, 169
Vedantasutra-75, 93, 140
Vedantika-31
Vedapurusa-204
Vedarthasangraha-64
Vedas-4, 15, 19, 23, 24, 42, 51, 55, 56, 76, 81, 84, 93, 105, 123, 145, 148, 199, 206, 207, 209, 213
Vedic-71, 79, 84
Vedic Ritualism-51
Videhamukti-11, 12, 13, 14, 31, 63, 115, 119, 127, 135, 137, 145, 150, 153, 157, 158, 165, 169, 170, 171, 172, 214
Vidyaranyasvami-20, 127, 128, 129, 131, 132, 133, 134, 136, 137, 138, 139, 140, 143, 144, 145, 146, 147, 149, 150, 151, 152, 153, 156, 157, 158, 160, 161, 162, 163, 165, 166, 169, 170, 172, 173, 176, 179, 180, 183, 184, 185, 191, 192, 193, 194, 195, 196, 198, 202, 203, 204, 205, 207, 208, 215
Vidyatirtha-130, 131, 144
Vijayanagara-131, 133
Vijnanavadins-27, 30
Virasaivism-79, 81, 82
Visesa-43
Visistadvaita-11, 63, 78, 128
Visnu-64, 76, 77, 78, 207
Vivarana-121, 123, 125, 126
Vivaranaprameyasangraha-140, 141
Vivartavada-70, 72

Vivek-179
Vivekacudamani-184, 214
Vyasa-98, 124, 156
Vyavaharika Satya-95, 111

Y

Yadavaprakasa-63
Yajurveda-34
Yajnavalkya-8, 9, 42, 146, 147, 171, 180
Yama-186
Yamunacarya-64
Yoga-50, 52, 78, 89, 90, 92, 103, 121, 151, 155, 165, 166, 172, 176, 182, 185, 186, 190, 192, 193, 200, 206, 207
Yogabhasya-186
Yogacara-27
Yoga Philosophy-54
Yogasutra-162
Yogavasistha-133, 134, 150, 151, 173
Yogic-29, 74
Yogic Advaita-143
Yogic Practices-54, 158

[illegible] the tongue, [illegible] just is Nirvana. Nirvana [illegible] extinguish. Nirvana consists in [illegible] sorrow [illegible] would not become unless the very cause of sorrow are [illegible] as is said in Dhammapada [illegible] down, is firm so long as [illegible] unless the feeders of [illegible] return again and again.

According to Buddha, overcoming [illegible] suppress some desires is not sufficient but complete eradication of them is necessary. It [illegible] forest, just not a tree only. Danger [illegible] when [illegible] undergrowth, then [illegible] the forest and tree.[12]

[illegible] Nirvana [illegible] not to indulge in any [illegible] weed out all the passions by means of [illegible] understanding. The Nirvana [illegible] a state of quiescence and eternal [illegible] consists in turning back upon all worldly pleasures and worldly attachments.

Buddhism believes that ignorance is the root cause of suffering. It is a [illegible] misconception of the world to which we [illegible] necessarily lead to a total disappointment and [illegible] pursues the right and avoids the wrong [illegible] This is possible only when one [illegible] right knowledge of things. As man's bondage is a creation of [illegible] based on knowledge, the Buddhists emphatically [illegible] that nirvana can be attained by proper knowledge [illegible] ignorance and illusion. [illegible] the absence of right knowledge of things. [illegible] of knowledge but it consists in wrong knowledge or distorted knowledge. It consists in understanding things as they are not in understanding them otherwise than what they really are.

The Buddhists not only recognised the efficacy of

12. Dhammapada, Max Müller (Tr.) 338 [illegible] 80, 283, p. 68